INFLAMMATORY BOWEL DISEASE

INFLAMMATORY BOWEL DISEASE

A Clinical Approach

Second Edition

Henry D. Janowitz, M.D.

New York   Oxford
OXFORD UNIVERSITY PRESS
1994

Oxford University Press

Oxford   New York   Toronto
Delhi   Bombay   Calcutta   Madras   Karachi
Kuala Lumpur   Singapore   Hong Kong   Tokyo
Nairobi   Dar es Salaam   Cape Town
Melbourne   Auckland   Madrid

and associated companies in
Berlin   Ibadan

Copyright © 1985, 1994 by Henry D. Janowitz

Published by Oxford University Press, Inc.,
200 Madison Avenue, New York, New York 10016

Oxford is a registered trademark of Oxford University Press

All rights reserved. No part of this publication may be reproduced,
stored in a retrieval system, or transmitted, in any form or by any means,
electronic, mechanical, photocopying, recording, or otherwise,
without the prior permission of Oxford University Press.

Library of Congress Cataloging-in-Publication Data
Janowitz, Henry D.
Inflammatory bowel disease : a clinical approach /
Henry D. Janowitz. — 2nd ed.
p.   cm.   Includes bibliographical references and index.
ISBN 0-19-507830-6
1. Inflammatory bowel diseases.   I. Title.
[DNLM: 1. Colitis, Ulcerative—diagnosis.
2. Colitis, Ulcerative—therapy.
3. Crohn Disease—diagnosis.
4. Crohn Disease—therapy.
WI 522 J34i 1994]
RC862.I53J36   1994
616.3'44—dc20   93-30353

2 4 6 8 9 7 5 3 1

Printed in the United States of America
on acid-free paper

For Adeline, again

A Brief Preface
and Some Acknowledgments

My introductory chapter in this book has tried to answer the question why I rewrote my personal view of inflammatory bowel disease at this time. But an earlier question needs to be addressed. At a time when not only new drugs pour forth in profusion, but multi-authored texts on inflammatory bowel disease are being published almost weekly, why another book at all? Especially at a time when information on all the varied aspects of IBD have become so highly specialized.

When so much remains controversial about the very concepts and the roles of medical and surgical therapy of these illnesses, I thought it useful to distill one person's experience of these disorders—my own. For many years, during each working day, I have struggled, along with my patients who suffer from these enigmatic and cruel diseases, in an effort to help them feel better. This volume is directed to the clinician and it is one clinician's appraisal. I respect my colleagues' views, but the material I present is based essentially on my own experience.

Its flaws are my own, but this volume has profited immensely from the careful reading and the blue pencil of Dr. David Sachar, the Burrill B. Crohn Professor of Medicine, my former fellow and now my chief as Head of the Division of Gastroenterology at The Mount Sinai Hospital, and from the detailed review of the manuscript by Dr. Asher Kornbluth, my former fellow, now Assistant Professor of Medicine at the same institution, and my valued colleague.

Dr. Daniel Maklansky, my radiological colleague for many years, gave me more than the radiographs for this volume and I would be remiss if I did not thank him for all that I have learned from him, besides radiology, through our daily discussions.

My editor, Joan Bossert, encouraged this project from the very beginning and helped make it coherent.

New York H. D. J.
April 1994

Contents

I *Introduction and Preliminary*
1. Introduction: Why Now? 3
2. Disease or Diseases? The Problems of Identity, 8
3. The Nature of the Beasts: A Speculative Chapter with Some Dogmatic Comments, 15
4. Theories of Etiology and Their Therapeutic Implications, 19

II *Clinical Presentations*
5. Modes of Clinical Presentation, 31

III *Diagnostic Modalities*
6. Notes on the Physical Examination, 51
7. The Essential Investigations, 54

IV *Problems in Diagnosis*
8. The Limits of Accuracy, 65
9. Making the Difficult Diagnosis, 68
10. Differential Diagnosis and Errors in Diagnosis, 72
11. The Obscure Differential Diagnosis, 80

V *Summing Up the Diagnostic Stance and Beyond*
12. Assessing the Patient's Clinical Status and the Degree of Impairment, 87
13. Prognosis: Predicting Outcomes and Reading the Future, 95

VI Medical Management

14. The Natural History of Inflammatory Bowel Disease: The Lessons of the Placebo, 107
15. Medical Management Programs: General Drug Considerations, 112
16. Supportive Supplemental Therapy: Nutritional Considerations, 144
17. The Role of Psychotherapy, 150
18. How Effective Are Our Current Drugs: Does Meta-analysis Help? 154

VII Surgical Management

19. The Decision for Surgery, 161
20. The Choice of Operations, 170
21. The Results of Surgical Therapies, 180
22. Postoperative Problems: Early and Late, 189
23. The Quality of Life and Inflammatory Bowel Disease, 200

VIII Special Problems

24. The Problem of Fistulas, 207
25. Pregnancy and Inflammatory Bowel Disease, 217
26. The Cancer Problem, 222
27. Associated Diseases, 230
28. Extraintestinal Manifestations, 235

A Personal Bibliography, 253

A Selected Annotated Bibliography, 265

An Album of Small Bowel Radiographs, 293

Index, 315

I

INTRODUCTION AND PRELIMINARY

1

Introduction

Why Now?

The last decade has witnessed a remarkable increase in valuable observations, in controlled therapeutic trials, and in the findings of clinical research and investigation in the area of inflammatory bowel disease. To one who, like myself, has lived through the last fifty years in this field, the abundance of new material seems all the more remarkable having taken place after a period of relative quiescence, one might even say of somnolence.

It is not merely the flood of publications in gastrointestinal journals throughout the world, and the new editions of multiauthored authoritative monographs on both sides of the Atlantic, the appearance of an encyclopedic text on the large bowel, to which many of us have contributed, and the literally hundreds of abstracts and poster presentations at American and European "digestive disease weeks" during this period. Not a month goes by without a regional, national, or international meeting, colloquium, conference, or symposium on IBD somewhere in the scientific world.

There are so many areas of important new information pouring forth that no doubt some interpretive synthesis would be useful. Judging from the reception of the earlier edition of this book and its reviews, I hope that this decade's internists, gastroenterologists, surgeons, intestinal intervention radiologists, and endoscopists, as well as gastrointestinal fellows and residents, will also benefit.

Almost at random, one could start with the effects of the wealth of basic and, at times, molecular information on the mediators of inflammation which have led to the profusion of therapeutic drugs; leukotrienes, interleukins, the free O_2 radicals have called forth their therapeutic antagonists.

The focus on the immunological behavior of the intestinal mucosa, beginning with the luminal epithelial cell, and including the intraepithelial lymphocyte and the intramucosal immume system, has served to unify our understanding of the responses of the intestinal and colonic mucosa to the ocean of antigens they are presented with daily and are awash with.

The classic IBD diseases—chronic ulcerative colitis (CUC) and Crohn's disease (CD)—remain the most important numerically. Yet newer studies reveal the importance of numerous microorganisms, heretofore considered "innocent bystanders," especially in the patient immuno-compromised with tumors, chemotherapy, or HIV viruses.

Clinically, there seems to be an emerging consensus of thought on the competing methods of evaluating the degrees of inflammatory activity going beyond simple clinical indices, the increasing use of leukocyte label–scanning techniques, extending to the measurement of mucosal levels of the very mediators of inflammation.

Therapeutically, a number of medications demand evaluation, spurred by the brilliant discovery of Truelove and associates that the active principal of sulfasalazine, the standard drug in this field since 1946, is the 5-ASA moiety. A plethora of preparations of this substance for oral and topical application (olsalazine, mesalazine, balsalizide) in a large number of increasingly available convenient forms (Asacol®, Dipentum®, and Rowasa® enemas) seems about to overwhelm the practitioner in their variety. Especially of interest is the recent observation that 5-ASA may be effective for maintenance therapy in Crohn's disease.

This needed evaluation is now possible since a number of workers, including those of our own group at The Mount Sinai Medical Center using meta-analytic methods, have calculated the therapeutic advantage of our current drugs in active treatment and maintenance trials of ulcerative colitis and Crohn's disease over placebo with recently developed mathematic techniques, which now can serve as standards of comparison for the newer compounds.

The use of immunosuppressive therapy for IBD beyond steroids (azathioprine and 6-mercaptopurine) is mature enough now to allow for realistic judgments regarding their place in the current management of active disease, and I believe that realistic projections can be made about the role of cyclosporine while we await the entrance of FK506 as it moves from transplantation surgery to IBD.

These judgments about our currently available medications are needed, especially, I believe, as candidate drugs wait in the wings. These include the poorly absorbed steroids (fluticasone proprionate or budesonide), leukotriene inhibitors, and immunological modulators, such as methotrexate and hydroxychloroquine, as well as dietary factors as the Omega 3 fatty acids.

The decade's newer surgical experiences already allow judgments to be made regarding their value in management, as well as their limitations and defects. Stricturoplasty for Crohn's disease instead of resection (as a bowel-saving operation) is now firmly established. Dilatation for stenosis of the intestine and colon remains a hope as yet unfulfilled.

The most valuable advance in the surgery of chronic ulcerative colitis is, of course, the rectal-sparing operation associated with mucosal stripping and the fashioning of an ileal pelvic pouch. Its cosmetic advantages make it an attractive alternative to the classical ileostomy and pancolectomy, but I believe its place requires considerable judgment with qualification in the face of the risk and development of "pouchitis." A serious inflammation of the pouch itself, this complication is now clearly established as not stemming from a confusion between patients with Crohn's disease with chronic ulcerative colitis, as our own and other studies have confirmed, but represents one human experimental model of IBD. My personal experience with this complication may be of interest. The revival of the ileoanal anastomosis in ulcerative colitis also requires some qualifying judgments regarding its risk, especially for the development of cancer.

Before the current form of colonoscopic surveillance for the development of cancer in chronic colitis hardens into dogma, the decade's experience, as reflected in my own experience, needs some reinterpretation, supported by the recent studies on the meaning of strictures in both CD and CUC, and especially on the limitations of mucosal biopsy.

The recognition of the increasing evidence regarding the mother's and fetus's tolerance for standard therapy for IBD has now been extended to immunological therapies as well.

Important evidence has been forthcoming from our own and other institutions on the magnitude of the cancer problem in both ulcerative colitis and Crohn's disease. Equally disturbing is the evidence that has been accumulating recently on the extraintestinal cancers and these disorders (lymphomatous and leukemic). New information has been accumulated regarding the whole field of extraintestinal manifestations of IBD. In addition to those of skin, eye, and joints related to activity in IBD and those that are due to intestinal pathophysiology (gallstones, oxylate and uric acid, renal stones, fat malabsorption, bile salt catharsis, vitamin B12 defects), we are now seeing a relationship between IBD and other acute autoimmune disorders. The striking treatment of the almost invariably fatal complication of renal amyloid with colchicine for this disease brings a breath of hope and relief.

The variety of options now open to clinicians for the treatment of the associated disease of sclerosing cholangitis: colchicine, ursodecholic acid alone or in combination, and methotrexate, especially in an effort to prevent the need for liver transplantation, are among the topics that deserve a reevaluation and that have been among the many fruitful areas of this last decade's research.

Introduction

But my main justification for revising this short book on nonspecific inflammatory bowel disease (CUC and CD) must remain my original intention of presenting only one perspective—my own—to offer a clinician's appraisal of these still cruel and enigmatic disorders. My aim is to put together what I have learned from nearly half a century's experience as a clinician and clinical investigator revised in the light of the last decade's private and public experience.

I have had the good luck to be involved with these patients through an exciting era of research at an institution and in a department deeply concerned with understanding their problems and dedicated to discovering how to manage their devastating effects. Ever since I met Burrill Crohn, Leon Ginzburg, and Gordon Oppenheimer when I arrived at Mount Sinai Hospital in New York in 1939, I have had the sense of participating in an exciting endeavor—which has been at the frontiers of both ignorance and uncertainty. Looking back, however, it is clear to see we have also contributed to the expanding body of knowledge we now have on IBD.

But as clinicians, and it is primarily to them that I address myself, we must continue to manage our patients as best we can despite the ambiguities of our current information and in the light of our own concepts. This small book can be considered as only one person's clinical appraisal.

2

Disease or Diseases?

The Problems of Identity

I still remain convinced that ulcerative colitis and Crohn's disease are separate and distinct entities; this conviction is more than a matter of faith. Yet some experienced observers believe that ulcerative colitis and Crohn's disease are separate ends of the same clinical spectrum. They could and do interpret the recognition that Crohn's disease can occur in the colon as supportive of their point of view.

In this context, it may be hard for the contemporary generation of clinical gastroenterologists to understand why it took so long to recognize such an obvious phenomenon as Crohn's colitis. The struggle to separate it from ulcerative colitis, which occupied, even preoccupied, clinical investigators during the period of 1932 to 1960, seems an episode of ancient history. Indeed, Ginzburg and Oppenheimer, in their first address to their surgical colleagues in 1933, pointed out that some forms of "terminal ileitis," their original designation of the entity "regional ileitis," which they first described with Crohn in 1932, did involve the large bowel as well; Wells had used the expression "Crohn's disease of the colon" in

1952. The modern era was clearly begun in 1960 with the classic paper by Lockhart-Mummery and Morson on the distinction of Crohn's disease (regional enteritis) of the large intestine from ulcerative colitis. (At my own institution, the pathologist Otani perpetuated the dogma that Crohn's disease always remained proximal to the ileocecal valve, and Burrill Crohn himself was among the last at Mount Sinai to accept the concept of Crohn's disease of the colon.)

Yet it should not disturb us that, even in the hands of pathologists skilled in these disorders, a portion of resected colons, as well as biopsies, resist classification and remain an *indeterminate* group. With increasing pathologic search, the number of these indeterminate colons continues to shrink and may not be more than 10 percent.

Further, the study of families reveals overlap between the two entities. McConnell has postulated that for Crohn's disease a person needs more of the genes responsible for susceptibility than is required for ulcerative colitis. Then again, in many studies of humoral and cellular immunology, a number of the demonstrated abnormal responses are shared by both disorders. Like tuberculosis and sarcoidosis, ulcerative colitis and Crohn's disease are shadow diseases. They seem to follow each other around.

To this we may respond that the very existence of an indeterminate group carries with it the implication that we are trying to force all types of colitis of unknown origin into too few categories. The indeterminate group may represent other as yet unrecognized inflammatory bowel diseases. Once all nonspecific colitis was considered ulcerative colitis; then there were two forms when Crohn's disease was separated out. Later, we had three since ischemic bowel disease was documented.

Then the floodgate was opened: infections with *Yersinia enterocolitica*, *Campylobacter jejuni*, and the pseudomembranous enterocolitis associated with *Clostridia difficile* toxin. Now, with the recognition of the role of *Cryptosporidia*, *Isospora belli*, and *Mycobacterium avium*—intracellular as well as cytomegalic virus in immunocompromised patients whether as the result of neoplasm, chemotherapeutic agents, and/or HIV infection—the list of precipitating agents has grown even longer. So the existence of an indeterminate group presents us with a challenge to further define this conglomeration rather than to put them all back again into the same "nonspecific" basket.

But the family overlap does not tell us that these inciting agents for ulcerative colitis and Crohn's disease are the same, but may only define their genetic susceptibility as witness the HLA B27 antigen that makes the subject open to the axial arthritis of Crohn's disease or Yersinial infection or Beçhet's disease. In this context, the recent studies of MacDermott and his colleagues are of great interest. These workers have shown recently that the spontaneous secretion of IgG antibodies by intestinal mononuclear cells showed clear-cut differences between ulcerative colitis and Crohn's disease, as well as from controls. Similar differences exist in the serum in the increased percentage and concentration of IgG, in ulcerative colitis from Crohn's, whereas patients with Crohn's have increased concentrations of IgG_2. While this observation may not have diagnostic value for clinical use, it certainly suggests fundamental differences in how the local intestinal cellular immune regulating system of these disorders reacts.

Crohn's disease and ulcerative colitis are often thought of as being in some sense *autoimmune* disorders if not immunologically mediated. It may not be a very strong point of their difference, but it is interesting that, while less than 2 percent of patients with Crohn's disease are associated with disorders currently considered of autoimmune origin, approximately 10 percent of patients with ulcerative colitis have been found to be so associated.

Even more important is the exciting discovery of the presence of a serum antineutrophilic cytoplasmic antibody (pANCA) in patients with inflammatory bowel disease, which differs in its perinuclear distribution from the classic cANCA of Wagner's granulomatosis. What is striking is that this serum marker is mainly associated with ulcerative colitis (50 to 80 percent in contrast to only about 1 percent in Crohn's disease). It appears to be present in patients also with sclerosing cholangitis and in a small percent of control sera.

Equally important is the fact that the antibody is independent of disease activity or extent and can be found up to 10 years following postcolectomy for ulcerative colitis and has a heterogeneous distribution in the relatives of ulcerative colitis patients with an increased frequency in relatives, specifically of ANCA-positive patients.

The complexity of the problem of identifying the immunoregulatory genes in inflammatory bowel disease is great; in addition to the ANCAs already discussed, workers have recently focused again on the HLA complex, the major histocompatibility complex. Rotter and his colleagues have noted the association of ulcerative colitis with HLA DR2, whereas Crohn's disease is associated with DR1 and DQW5. Study of these markers raises the real possibility that there may be more than one form of ulcerative colitis and Crohn's disease, maybe even combinations of both. (I recall Burrill Crohn's remark that he saw no reason why a patient might not have features of both of these diseases.)

So it is, in the end, the natural history and pathologic substrate of these two disorders that at present constitutes for me the main support for their distinctiveness. The repeatedly emphasized, well-known clinical features: relative sparing of the rectum, greater number of abdominal masses, increased instance of perirectal and other fistula (intra- and extra-abdominal), and distribution throughout the entire gastrointestinal tract of Crohn's disease serve to distinguish that disease from ulcerative colitis, combined with its greater tendency to form fibrotic stenosis and the curious tendency to "metastasize" to the skin remote from the gastrointestinal tract as well.

It is routine at present for the pathologist receiving a colonic biopsy or resected colon to find the presence of granulomas as being virtually pathognomonic of the diagnosis of Crohn's disease, although, in most series, they are present in only 70 percent of cases even with particular study of the colonic and regional lymph nodes. The only 2 to 3 percent of granulomas that have been reported in patients "classical" for ulcerative colitis does point to some important differences between the two disorders, perhaps only in the rate or way the local immunocompetent cells process some unknown antigen or antigens. The fact that we have seen the development of mucosal granulomatous systems in two of 15 severe refractory patients with pouchitis whose restudied colons were clearly ulcerative colitis must give us some pause in making the sarcoid lesion an absolute marker of Crohn's disease.

In considering the similarities and differences between ulcerative colitis and Crohn's disease, it is pertinent to look at the respective response to surgical intervention. It has long been accepted that total colectomy "cures" ulcerative colitis with all lesions arising in the small bowel ascribed to mechanical problems at the cutaneous stoma, whereas the relentless rate of recurrence of Crohn's disease is generally not affected by any surgical resection. The results of the creation of intestinal pouch reservoirs by the current procedure of the pelvic pouch and the rectal-saving maneuvers in ulcerative colitis are pertinent to this discussion.

Typical inflammation in the pouch, "pouchitis," has been reported in about 30 percent of ileo-reservoirs and is conventionally ascribed to ileal stasis and bacterial overgrowth. But it becomes obvious that there must be other factors when one considers that pouchitis rarely occurs in individuals being operated on for familial polyposis. The majority of inflamed pouches respond to antibiotics, especially metronidazole therapy, yet severe inflammation in some with endoscopically atypical features are relatively refractory to our current therapy. This refractory form is conveniently ascribed to misdiagnosis of the colonic disease as ulcerative colitis in individuals who are believed to have suffered from overlooked Crohn's disease of the colon. A review of some 25 such individuals in our group with an independent, blinded reexamination of the resected colons refutes the conventional wisdom that their pouchitis was due to the diagnosis of Crohn's disease having been missed in the resected colons, as does the recent observation from our center and others that ANCA-positivity is at least as high among patients with refractory pouchitis (if not higher) as among the ulcerative colitis population as a whole.

In connection with the family overlap of ulcerative colitis with Crohn's disease just discussed, the study of twins in Scandinavia among monozygotic siblings shows that concordance of the disease in the twins is very much more likely to occur with Crohn's than with ulcerative colitis, a further small point regarding their fundamental differences.

An interesting, but probably not an overwhelming, point in the argument regarding identity is the idea that, if they were really the same diseases, they should occur together in some patients whose tissue responses would show conponents of both disorders as now characterized by informed pathologists. These patients should have more shared clinical traits. However, few well-documented accounts of their coexistence exist, but we should not deny a priori the possibility of other disorders coexisting in the same individuals.

The epidemiology of inflammatory bowel disease considered especially with time trends may also be pertinent in this attempt to distinguish between ulcerative colitis and Crohn's disease. Despite the inherent problems of measuring the incidents of these disorders in recent times, especially after 1960 when Crohn's disease of the colon was sharply defined, two generalizations appear to me to hold. There has been a rising instance of Crohn's disease prior to 1980; the bulk of data generally shows a more steady, continuing incidence trend in ulcerative colitis. This, on the surface, seems to emphasize their difference. Yet, McConnell's hypothesis of a shared genetic background with two different environmental factors (one stable and one rising) could theoretically be operative.

I believe that here we are dealing with "the substance of things unseen." We must continue to distinguish these disorders until their etiologies are well established. We need to do this both to manage our patients rationally and to make progress in understanding their clinical behavior. The history of clinical investigation favors "splitters" rather than "lumpers."

After I had written the preceding material, I noticed that the problem of the identity or the difference between Crohn's disease and ulcerative colitis (one disease or two?) is posed most clearly in the titles of two authoritative current monographs. Drs. Joseph Kirsner and Roy S. Shorter call the third edition of their multiauthored text *Inflammatory Bowel Disease,* while Drs. R. N. Allen, M. R. P. Keighley, J. Alexander-Williams, and Clifford Hawkins call the second edition of their London-based book *Inflammatory Bowel Diseases.* Like many other authors, I find myself in both books, but I am not in both camps. I believe we shall make more progress if we continue to keep these disorders separate, and report and remember our experience separately.

3

The Nature of the Beasts

A Speculative Chapter with Some Dogmatic Comments

Knowing that the causes of ulcerative colitis and Crohn's disease are unknown, readers who find speculation unprofitable may want to skip this chapter and move on to the more practical details of clinical management. But, as we all struggle with the daily care of our patients with these disorders, we tend to have some vaguely formulated, perhaps incoherent, concepts of the natures of these disorders. From time to time, in the midst of the details of daily management, we must ask ourselves what kind of diseases these are.

What would an experienced clinician combining the innocent eye of the child with the sophisticated skepticism of a clinical investigator see as he or she looked at these diseases without preconceptions? Perhaps the following could be seen:

1. *Essentially a group of intestinal disorders.* Although we know that Crohn's disease can spill over from the gut into the larynx, around to the skin from a fistula, become slighltly disseminated from the intestine to the liver, bones, amd more distantly to the skin, primarily these are diseases of alimentary tissues: mucosal ulcera-

tions and friability, mucosal canker sores (aphthous ulcers) of the mouth and gut, surface ulcerations of the large and small bowel, and superficial patches of necrosis on histological examination.

2. *Diseases of the distal intestine.* Ulcerative colitis is a disease of the colon alone, although of course Crohn's disease may occur throughout the entire tract. Ulcerative colitis is confined to the colon and, if the colon is extirpated, the ulcerative colitis is "cured."

Crohn's disease, in spite of its ability to appear anywhere in the intestinal tract, is mainly a distal disease occurring most frequently in the ileum and the ileocecal area and more interestingly occurring behind sphincters: the ileocecal and the rectal.

3. In Crohn's disease, I have for a very long time had the intuitive feeling that whatever the triggering agent or antigens, the exciting material is in the intestinal lumen and moves downstream being concentrated or held up at the distal sphincters. This theory is consistent with the distal nature of the disorder in its ileocecal and perianal locations, together with its relative rarity in the upper gut.

Further, the observation that bypass operation in the ileum for regional ileitis, with exclusion of the bypassed loop from the intestinal stream, led to healing and fibrosis during the pioneering era of surgery for Crohn's disease has impressed me very much. Some recent ingenious surgical experiments abroad have reemphasized the importance of the intraluminal contents when in continuity and the protection of the bowel from inflammatory changes when it is excluded from the fecal stream in Crohn's disease.

Long surgical experience in ulcerative colitis had repeatedly shown that a diverting ileostomy does not improve the colonic disease, and further that subtotal colectomy leaving disease behind in a Hartman pouch does not improve the local rectosigmoid colitis.

4. Crohn's disease and ulcerative colitis manifest a lifelong tendency to recur, and in the same anatomical site, more often in Crohn's disease than in ulcerative colitis as if the susceptibility of these areas were due to a specific local defect; further, the very striking characteristic recurrences at anastomotic sites in Crohn's disease are on the proximal side of the anastomosis, especially in ileitis.

5. Diseases that occur frequently in family clusters with members of the same family sometimes having both forms.

6. To these observations we ought to add some historical material since our hypothetical observers would be familiar with older literature. Inflammatory bowel diseases are relatively new disorders—that is, the "nonspecific" ones. The information is more obscure in the question of ulcerative colitis but it was surely separated from other specific infections involving the colon no more than a hundred years ago. Crohn's disease, it strikes me, is clearly a new disease. It is difficult to believe that the great physicians of the European nineteenth-century school, Virchow and Connheim, in Berlin and Vienna, who performed complete autopsies and who did routine small bowel run-through in all their autopsies, would not have recognized Crohn's disease. They were well aware of tuberculosis and I do not think they would have confused it. After all, the first good description of what was to be Crohn's disease was by the Scotch surgeon Dalziel, in 1913, and in a small group of cases by Crohn, Ginzburg, and Oppenheimer in New York at The Mount Sinai Hospital in 1932, just about the same time that Arnold Bargen at the Mayo Clinic was collecting his cases.

Further, from a historical perspective, Crohn's disease has certainly been considered to behave like an infectious disease from day one. Its general characteristics resemble such chronic and progressive infections, for instance, as tuberculosis or actinomycosis. The search for the infecting agent had preoccupied investigators from time to time and is always under vigorous pursuit. In this context, we should remember that improved microbiological techniques have slowly allowed us to separate other specific causes from the welter of nonspecific inflammatory bowel disease. These include *Yersinia enterocolitica,* campylobacter infection, which mimics ulcerative colitis, and brilliant recognition that the toxin of *Clostridium difficile* can produce a pseudomembranous enterocolitis as well as the diarrhea associated with antibiotics.

Now we are deluged with information regarding a whole host of intestinal organisms incurring in immunocompromised individuals which in turn has opened up new vistas.

Everyday it seems that new organisms or old ones are reported from the intestinal tract so that gastroenterologists can hardly keep up with them. What am I to make of the report of *Aeromas hydrophilus* in the stool recently reported in a patient seen in consultation who had what looked like straighforward ulcerative colitis? There is this continuing suspicion that in Crohn's disease we are dealing with a particularly fastidious organ, which, so far, has eluded our techniques.

In this perspective, we must include the fascinating series of clinical phenomena that occur in these patients, changes that are different from the intestinal disease—conjunctivitis, choroiditis, erythema nodosum, pyoderma gangrenosum, peripheral arthalgia and arthritis, now considered due to immune complexes, ankylosing spondylitis, and sacroiliitis, to say nothing of the nephrotic syndrome with amyloid deposition.

And then we may add that in some individuals with these disorders, there is a curious characteristic pathologic tissue reaction—the lymphoid granuloma—either in the gut or the regional lymph nodes occurring with varying degrees of hardness and softness. In some of these patients, amyloid deposition also develops.

Perhaps one might consider such an elastic formulation as this: in a susceptible population with some altered immunoregulatory genes, one or several agents (antigens) present in the intestinal contents are moving downstream to find or create a disruption in the intestinal mucosa activating the local epithelial immune system to release a host of mediators of both acute and chronic inflammation. And at the same time there are released or absorbed other antigens which lead to the deposition of immune complexes at distal sites, such as the eyes, skin, or joints. I believe we could all unite under such a broad banner.

4

Theories of Etiology and Their Therapeutic Implications

Granted that the etiology or etiologies of ulcerative colitis and Crohn's disease are still unknown, yet the current climate of possible etiologies must and does influence our therapeutic approaches to our patients with these disorders. Controlled clinical trials are not the only determinants of our therapeutic behavior. So I propose in this chapter to consider the impact of our current theories of causation on our treatment procedures and to comment as well on any etiologic inferences we might draw from our current drugs.

Epidemiological approaches have shed little light and at present few therapeutic considerations have resulted from these approaches.

Patients with ulcerative colitis are mainly nonsmokers and ex-smokers show an increased risk. These patients, however, have told us for a long time that they date the onset of their problems to stopping smoking. Patients with Crohn's disease are more likely to be smokers. In my experience, few patients with Crohn's disease date the onset of their illness to stopping smoking. Whatever the mechanisms involved, promotion of Crohn's disease or protection against ulcerative colitis, I think we all agree that the risks of cigarette smoking justify our prohibition of smoking in all patients. However, in connection with the increased number of smokers with Crohn's disease, it is interesting that recent British workers, Pounder and his group, have emphasized the possible role of local vascular pathology in the pathogenesis of this disease.

Contagion as a factor has had little support in the past since few space-time clusters have been reported, although all experienced observers have noted, on rare occasion, husbands and wives who have contracted the disease. I have an interesting "cluster" of three male law students who, shortly after leaving their university and four years of close living contact, came down with inflammatory bowel disease—two with Crohn's and one with ulcerative colitis. I also have several families in which one spouse developed the same inflammatory bowel disease as the other after marriage, but I have thought, as have others, that this is a rarity. The recent study by my colleague, Daniel Present, and his co-workers, has raised considerable interest in this context, since, in their 19 couples, where both husbands and wives were affected, five had symptoms before marriage, whereas, of the remaining 14, one partner had the disease before marriage in seven couples, and, of the remaining seven couples, both developed the disease after marriage; in these, diseases were discordant in six. Surely no one would or could discourage any couple with inflammatory bowel disease from getting married. But the possible 30 to 50 percent risk of Crohn's disease in the offspring of two patients with Crohn's disease might give one pause before encouraging such afflicted pairs to have children.

Genetic Factors

All who treat or study inflammatory bowel disease have been impressed from the earlier reports of this century by the increased prevalence of this disorder among the families and relatives of ulcerative colitis and Crohn's disease patients, the association extending even into second-degree relatives.

What can we say at present about the genetic risk of being born into a family with inflammatory bowel disease? Perhaps up to 7 percent of patients with IBD will have a child with the disease, and perhaps up to 9 percent of patients with IBD will have a sibling with the disease. Further, Crohn's and ulcerative colitis occur in greater numbers than can be expected in the same family tree.

In this genetic context, there is no question that Jews appear to be at highest risk for inflammatory bowel disease, probably at least three times higher than the non-Jewish population, both for Crohn's and for ulcerative colitis. At risk especially are Jews of Middle European origin and among the Ashkenazi rather than the Sephardic Jews.

Twin study linkages emphasize the greater risk of genetic factors in Crohn's disease than in ulcerative colitis. If one is a monozygotic, that is an identical twin with Crohn's disease, the other twin will probably develop Crohn's disease in 65 to 70 percent of the cases. If one is a dizygotic twin with Crohn's disease, the other twin will develop Crohn's disease in 8 percent. If one is a monozygotic twin with ulcerative colitis, 20 percent of the unaffected twins develop the disease versus 0 percent developing the disease if one is a dizygotic twin with ulcerative colitis. The implication here is that since, even among monozygotic twins with Crohn's disease, there is concordance for Crohn's disease in only 65 to 70 percent of the cases rather than 100 percent, as it would be if it were a purely genetic disease, where there is a genetic susceptibility to an environmental agent or agents. Also genetic factors are responsible for family aggregations because of the higher risk of developing the disease in monozygotic identical twins versus nonidentical twins. Lastly, Crohn's disease is more "genetic" than ulcerative colitis. Interestingly enough, smoking was more common in monozygotic twins with Crohn's disease than in monozygotic pairs with ulcerative colitis, but the smoking could not explain discordance in both diseases. (Indeed, identical inheritance and similar smoking factors apparently are not enough to guarantee concordance disease.)

In the context of genetic susceptibility, the current actively investigated area of increased permeability of the intestine in unaffected relatives of patients with Crohn's disease may contribute a link to the role of diet in susceptibility in this disease, although this has now been challenged.

These observations seem so well established that McConnell as recently as 1990 stated that the stage had been reached in the study of the genetics of inflammatory bowel disease that this information "can be used for genetic counseling"! We may and probably should share with our patients what secure or firm information we have regarding the hereditary factors of these disorders, but we certainly should not, and probably could not, in my experience, intervene in the marriage decision.

Dietary Factors

Stimulated by the discovery that gluten intake plays a fundamental role in the etiology of celiac sprue, investigators have pursued the question of a dietary factor (excessive or deficient) as a possible etiologic factor in the inflammatory bowel disease groups. The difficulties of measuring the diets patients ate before their coming down with illness are great. A sugar-rich diet seems to be a consistent finding in Crohn's disease patients. This does not seem to be related to the effect of smoking; smoking and sugar consumption seem to be separate but interactive. For the present, this "fact" of higher sugar in the diet does not seem to have had any implication for the management of patients. Perhaps what is needed is an ongoing study of the reduced role of sugar in the diet on the maintenance of remission in patients with Crohn's disease following the induction of remission after surgery.

So patients remain dissatisfied that they cannot pinpoint any errors in their previous diet which, when corrected, could improve their present clinical condition. (Current dietary therapy will be considered in the second half of this chapter.)

Infectious Agents

Mycobacteria

From the very first description of Crohn's disease, an association with tuberculosis and related mycobacterial infection has been suggested and diligently looked for. Dalziel, in 1913, saw the histologic resemblance between his "chronic intestinal enteritis" and the mycobacterial disease of cattle in the small bowel called Johne's disease. Crohn, Ginzburg, and Oppenheimer (1932) suspected that their "regional enteritis" might be of tubercular origin, but failed to find the organism by culture and guinea pig inoculation, however the search has continued. *Mycobacterium kansasii* and *Mycobacterium paratuberculosis* have been isolated on rare occasions in patients with Crohn's disease, but the bulk of other attempts to conform these findings have so far not been encouraging. Just as a number of organisms were recovered from the tissues of Whipple's disease before its recent and mysterious disappearance from the gastrointestinal tract, although a specific organism has recently been discovered, so Crohn's disease may also be the result of a variety of organisms: some mycobacterial, some cell wall–deficient forms as had been suggested by Parent and Mitchell (1976–1978). As one could easily predict, the workers who isolated *M. kansasii* and *M. paratuberculosis* would obviously utilize mycobacterial therapy. Double therapy with streptomycin and rifaputin or quadruple therapy with rifampicin, ethambutol, isonizid and either pyrazinamide or clofazamine has been applied to Crohn's disease in open studies with reported good effects, but the majority of clinical investigators who have pursued double-blinded placebo-controlled trials of these combinations of antituberculosis drugs in Crohn's disease have failed to confirm their value. The recent identification of the causative organism of Whipple's disease by the polymerase chain reaction gives rise to the hope that this approach may soon give us an answer or answers in Crohn's disease and possibly settle the role of atypical mycobacteria.

Anerobic bacteria have been looked at from an etiological point of view. There is a reduction in obligate anerobes and an increase in facultative anerobes in both active ulcerative colitis and Crohn's disease, greatest in patients with stasis and fistula formation, and this may play a secondary role in etiopathogenesis. These findings obviously set the stage for metronidazole therapy. The open and controlled clinical trials in Crohn's disease show some favorable responses to metronidazole, but currently most studies show little response in ulcerative colitis. Even in Crohn's disease, its ameliorating effect may not be due to its direct effect on anerobic organisms, but to some nonspecific methods by way of reducing bacterial overgrowth and improving ileal function and bile reabsorption, as well as possible immunomodulating effects.

Transmission studies stimulated by Mitchell and Rees, who produced granulomas in the footpads of mice with filtrates of intestinal homogenates of Crohn's disease patients' tissues, the only type of etiologic research I personally have been involved in, have fallen into disrepute. Yet the granulomas, which a considerable number of workers reproduced but not invariably, are probably not all artifactual and may not have transferred a replicating agent, but probably did transfer some mediator or mediators of inflammation, which could induce local granuloma formation.

Mediators of Inflammation

While the search for the activating event in inflammatory bowel disease goes on full speed, a large segment of current basic and clinical research is centered on the cellular mediators of inflammation in the intestine itself which offers the opportunity of therapeutic intervention.

The pathway of arachidonic acid metabolism, which release prostaglandins via the cyclooxygenase route and the lipooxygenase route, which in turn releases leukotrienes (LTB4, C4, C5, E4), as well as the phospholipases, which release the platelet activating factor, are well documented as being involved in the local inflammatory action in both ulcerative colitis and Crohn's disease.

Along with this mechanism, macrophages of the lamina appropria, as well as the recently described antigen-presenting functions of the intestinal epithelial cell itself, release a cascade of immune responses locally in the gut wall with the production of cytokinins, such as interleukin 1 and 6, tumor necrosis factor, and elevated B-cell activity, including enhanced Ig secretion, especially IgG, and perhaps the super-oxygen radicals, which are more a neutrophil and macrophage product than a B-cell activity per se.

In this area, the empirical therapeutic response of corticosteroids and sulfasalazine and its active principle (5-ASA) preceded the study of their tissue mechanisms involved in the inhibition of the inflammatory process. Corticosteroids by the induction of lipomodulin lead to an inhibition of a major step in the arachidonic acid pathway while sulfasalazine and 5-ASA inhibit both the cyclooxygenase and the lipooxygenase pathways and 5-ASA inhibits the release of the potent inflammatory mediator platelet activating factor.

The therapeutic implications of these findings clearly indicate that full scale research will continue to be devoted to the suppression of these mediators for the treatment of inflammatory bowel disease with emphasis already on steroids, which are cleared rapidly from the plasma, such as budenamide, interleukin receptor inhibitors, leukotriene inhibitors, including the trials of Omega 3 fatty acids, Plaquinal®, and methotrexate. This suggests that these inhibitors are modulators of the immunoinflammatory chain and might be more effective if exhibited simultaneously rather than successively as at present.

Immunologic Etiologies

The clinical effectiveness of corticosteroid therapy on inflammatory bowel diseases and their extraintestinal manifestations in the eye, skin and joints, along with the observation in 1963 by Perlmann and Broberger that peripheral white blood cells of individuals with ulcerative colitis could kill epithelial cells from fecal gut cultures, as well as the failure of a generation of investigators to find an infectious agent, has directed most current research in the etiology and etiopathology of these diseases to the search for immunological mechanisms.

The bulk of these researches have recently been adequately reviewed elsewhere together with the association of inflammatory bowel disease with autoimmune diseases.

There seems to be no defect in systemic immunity, so the active research is directed at present toward the study of defects in the gut wall immune system and its reactivity. Since the distal small bowel and the colon are awash in a sea of possible antigens (of dietary, microbial, or intestinal cellular origin), it is postulated that the inflammatory reaction of the gut is the result of either an appropriate response to an unusual organism, such as a mycobacterium or other microorganism, or an abnormal response to a common antigen.

Most intriguing is the concept that, while the normal gut can significantly depress the inflammatory response to the stream of fecal and intestinal antigens, the intestine of ulcerative colitis and Crohn's disease patients is unable to shut off this activation. Recent studies from our own institution demonstrate that the intestinal epithelial cells of normal subjects can act as antigen-presenting cells and selectively activate nonspecific suppressor T-cells (CD8 +). On the other hand, intestinal epithelial cells from inflamed and normal areas of patients sick with ulcerative colitis and Crohn's disease preferentially activate helper T-cells (CD4 +) and fail to activate appropriately the suppressor cell population. These studies suggest that the immunological defect in inflammatory bowel disease may reside in the intestinal cell itself.

Equally intriguing in this context is the finding of diminished interleukin-2 production by the lamina propria mononuclear cells of inflammatory bowel disease in both Crohn's and ulcerative colitis since this lymphokine has an essential role in regulating immune function. This abnormality of interleukin production of intestinal CD4 + T-cells appears to be due to a defect in the cells themselves.

While the therapeutic success of immunosuppression by corticosteroids and more recently by either azathioprine or 6-mercaptopurine encouraged the search for immunologic defects, these studies have influenced a search for more specific T-cell inhibitors, such as cyclosporine and the trial of hydroxychloroquine an antigen-presenting cell-mediated activating mechanism, as an inhibitor of CD4 + helper cells. This line of evidence also suggests that a diversity of factors might trigger the inappropriate local immunological response to injury, but does not solve the problems of the respective intestinal localization in these disorders, especially the problem of fistulization in Crohn's disease. Is the immunological defect in ulcerative colitis confined only to the colon in contrast to the wider distribution in Crohn's disease?

Equally intriguing from an etiologic point of view is the meaning of the demonstrated value of maintenance therapy in ulcerative colitis by sulfasalazine and more recently of the 5-ASA compounds in Crohn's disease. These facts raise the question of whether we are merely suppressing mini-flareups or are actively influencing the activating mechanism or possibly even both. (If peptic ulcer disease is a lifetime disease, what is the basic mechanism of maintenance therapy by histamine 2 blockers or *Helicobacter pylori* eradication?)

It is equally important from a *therapeutic point of view* that what we classify as ulcerative colitis and Crohn's disease are each possibly more likely to be more than one disease. This not unlikely conclusion seems to follow from the observation that HLA DR2 is associated more with ulcerative colitis and HLA DR1 or DW5 with Crohn's disease, whereas the presence of antineutrophilic cytoplasmic antibodies (ANCAs) is associated overwhelmingly more with ulcerative colitis than with Crohn's disease. Subsets of these associations also suggest heterogenecity within each clinical category.

II

CLINICAL PRESENTATIONS

5

Modes of Clinical Presentation

While there is considerable overlap in the way in which these diseases present themselves to the clinician, there are characteristic differences based on the differences in the pathologic process in ulcerative colitis and Crohn's disease.

Varieties of Ulcerative Colitis

Ulcerative colitis, essentially a colonic mucosal epithelial disease with attendant alterations in colonic motility and usually a distal colonic disorder, presents itself in a rather stereotyped manner varying mainly in the degree and intensity of rectal bleeding with the increased or diminished motor activity of diarrhea or constipation and valuable amounts of systemic symptoms.

Proctitis

Rectal ulcerative inflammatory disease is undoubtedly the commonest variety of ulcerative colitis and, since we are all well aware of the ominous sound of "ulcerative colitis" to our patients, we reassure them that the inflammation tends to remain localized to the rectum except in some small percentage of individuals, citing the number we summon up from a scanty literature. Indeed, we tend to consider this variant as a most distant cousin of ulcerative colitis in the family of inflammatory bowel diseases, much to the relief of our patients.

What I would stress here is that, in contrast to the textbook account, ulcerative proctitis is a stubborn disease: difficult to treat and frequent in its recurrences. Of course, it rarely is associated with the systemic symptoms of fever, loss of appetite sufficient to cause weight loss, and general malaise. Indeed, if such systemic features are present, I assume that the disease is present beyond the rectum, even beyond the sigmoid colon, extending into the left side of the colon.

I have never seen ulcerative proctitis associated with a rectal vaginal fistula or a urinary bladder fistula nor with any of the usual extraintestinal manifestations in skin, eye, or joints.

My complacency in the presence of this localized proctitis has been undermined by clinical experiences in seeing patients in whom the localized disease, under observation and local intensive therapy, can and did extend, occurrences I have observed particularly among younger patients.

In another quite common variant, the rectal bleeding, often associated with blood-stained mucus, presents itself with tenesmus and difficulty in the rectum's evacuation of its contents. There may be considerable local soreness and even some poorly localized lower back pain or aches.

I have found no particular characteristic age distribution, but my impression is that proctitis is a disease of the adult rather than that of the teenager and women seem to predominate slightly. It may be the first incidence in the inflammatory bowel history of the individual or the recurrence of such an incident after periods of short or quite long quiescence.

It does not surprise me that younger people—teenagers and some in their twenties—tolerate low-grade intermittent rectal bleeding and even an increase in the liquidity of their stools and in their number for long periods of time before informing their parents or seeking medical advice. This is especially so in the group having their first episode while away at college. What is more puzzling, is the number of mature adults who tolerate a change in bowel habits, which includes bleeding, for equally long periods of time despite all the public writing and warning that this change may indicate a cancer of the rectum or colon. Some of the adults have been reassured in the past that their rectal bleeding was due to associated hemorrhoids, but they have not had a sigmoidoscopic examination at the time of bleeding.

Left-Sided and Universal Ulcerative Colitis

Loss of appetite, loss of weight, fever, and more intense left-sided, lower abdominal pain are not characteristic of proctitis. When systemic features, just as those outlined, supervene, we are dealing with either a proctitis that has extended or a more acute form of proctosigmoiditis or left-sided ulcerative colitis. These patients are clearly sick and disabled and they do not tolerate their symptoms for very long before consulting a physician.

Universal ulcerative colitis presents with an increase in the intensity of the symptoms as described for proctosigmoiditis or left-sided ulcerative colitis in the more florid forms: high fever, loss of appetite and weight, marked abdominal pain and profuse diarrhea, and the characteristic hallmark of marked blood in the stool and often free rectal bleeding unassociated with bowel movements. However, this is not always the case. Patients can present with radiographic and colonoscopic evidence of involvement of the entire colon with the much less dramatic symptoms of a low-grade yet continuously active disease process. The onset of universal ulcerative colitis may be insidious with slowly increasing intensity of symptoms or may be explosive at the onset beginning precipitously with an acute dramatic flourish without much prologue.

While proctitis may have as its initial symptom *rectal bleeding* alone without any other initial symptoms and without any disturbance in bowel problem, bleeding is not usually profuse, although the patient is disturbed by the blood staining the toilet water. Massive hemorrhage can occur in a more extensive variety of ulcerative colitis, alarming in intensity and magnitude, but rarely as the initial presentation. Massive bleeding is more likely to occur in my experience in the course of known ulcerative colitis and most often in the setting of increased disease activity, although not necessarily fulminant in nature.

In contrast to Crohn's disease, *acute mechanical obstruction* rarely occurs in ulcerative colitis. Although the development of cancer in long-standing ulcerative colitis of the universal variety does lead to increasing obstruction, I have not seen a patient with complete colonic obstruction. However, on one occasion I observed a patient who felt almost complete splenic flexure obstruction in the presence of known ulcerative colitis from a tangle of a mass of pseudopolyps so thick that the colonoscope could not penetrate it.

Toxic dilation of the colon deserves separate treatment and receives it later on. This variety can occur very early in the course of the first episode of the disease, which certainly has become fulminant, and can occur in Crohn's disease, as well as ulcerative colitis. However, this form of presentation clearly has been occurring much less frequently at present in my own and other observers experience. This deduction I am inclined to ascribe to better and earlier medical management of acute colitis, avoidance of narcotics, and the performance of unnecessary barium enemas.

Free perforation independent of toxic dilation is a very rare presentation of ulcerative colitis. Walled-off perforation, by contrast, can occur frequently in the course of any acute episode of ulcerative colitis and may occur in the first one signaled by a sudden new pain, especially in the left lower quadrant, which the patient may recollect later on. Because of the presence of true visceral tenderness in ulcerative colitis, we tend I think to diagnose peritonitis and walled-off perforation more frequently than they occur.

The problems of extraintestinal manifestations in inflammatory bowel disease will receive ample discussion further on in this text in the section on special problems (Chapter 28), but their incidence in the initial presentation of ulcerative colitis is of interest. Our studies

Modes of Clinical Presentation

have fairly well demonstrated that those which are related to clinical activity (eye, skin, peripheral joints) appear to be dependent on the involvement primarily of the colon and its degree of inflammation. As I have already mentioned, localized proctitis is almost never associated with these extraintestinal manifestations. It is in left-sided and universal colitis that they occur. Iritis and/or conjunctivitis may either precede the acute onset by weeks, even months, or be coincident in time with the more obvious clinical onset. Erythema nodosum may appear in its first crop at the very onset of the disease or even a short time before. While pyoderma gangrenosum has been known to herald impending activity of severe ulcerative colitis, this is not usual. Patients and their bowels need to be sick for a long period of time for true pyoderma to develop. Arthralgias and fleeting pains in joints and joint aches are common at the onset of left-sided and universal colitis, but truly inflamed red, hot swollen painful peripheral joints are rare at the very onset of ulcerative colitis, whereas low-grade arthropathy in Crohn's disease, especially the "reactive arthritis" of ileitis, may precede the clinical symptoms of bowel disease.

Precipitating Factors

In my experience there are few obvious precipitating factors in the onset of ulcerative colitis in its various forms. Other systemic diseases are not usually present, especially in the younger population that contains the bulk of our inflammatory bowel disease patients. *Stopping cigarette smoking,* a dramatic and an easily dated change in behavior, has become a more frequent complaint, especially since we are now paying more attention to our patients' accounts. I can recall patients reciting the same story over many past years. Respiratory infections, especially episodes called "the flu" by patients, seem chance factors. Yet the latter may even represent the symptoms of the incubation period of the ulcerative colitis itself. On some occasions, single members of a family or group who have suffered from an acute diarrheal disorder ascribed to a particular meal may present with the persistence of symptoms which now

bleeding. Stools and duodenal drainage in these
...en in my experience have not turned up a parasite or
... to account for the original episode of the colitic
...ne cannot disregard the reports of some members of
an epidemic outbreak of amoebic or salmonella infections being left
with resultant chronic inflammatory bowel disease.

If a patient has been treated with an antibiotic for some intercurrent infection or the acute gastroenteritis ascribed to a particular meal, it is conventional to look for *C. difficile* toxin in the stool and to treat it if found. In my experience, this rarely turns out to be the case, although any antibiotic taken in the prior six months can lead to the presence of the toxin. However, longer experience has not convinced me that *C. difficile* toxin plays any important role in precipitating recurrence of inflammatory bowel disease. I think we should continue to look for it, but I do so with diminishing enthusiasm or hope that I will find it present or meaningful. Even when the toxin is present, the course of inflammatory bowel disease and its severity seems to bear no relation to its presence. Some few patients, even in remission, continue to harbor the toxin in their stool.

More recently, we have all become aware that the use of such nonsteroidal anti-inflammatory drugs as indomethacin can cause a flare-up of ulcerative colitis. This holds true of other members of the prostaglandin inhibitors.

Psychosomatic Factors as Precipitants

Insofar as we attempt to practice gastroenterology as a scientific discipline, we want and need hard data and we have little regard for the influence of emotional factors inducing ulcerative colitis or favoring recurrence of known or established disease. Our patients, however, and their families know that "nerves cause colitis." My own experience leads me to believe that there is more psychogenicity in the background of ulcerative colitis patients than in those with Crohn's disease, at least before the clinical onset of their inflammatory disorders. All experienced physicians in this century, long before the emphasis on "psychosomatic" or "holistic" medicine, have recognized the importance of emotional turmoil in the course of ulcerative colitis. My own impression is that a single, traumatic event is less frequently elicited in careful history-taking than are more sustained unresolved emotional pressures.

The Protean Variations in Crohn's Disease

Classic Presentations

By now the typical clinical presentations of Crohn's disease are well recognized. The most common is the younger patient (adolescent through young adults), who complains of rather sudden onset of abdominal pain, usually in the right lower quadrant, with low-grade fever, perhaps a slight increase in the number of bowel movements or slight softening of their consistency. Usually suspected of having an acute appendicitis, the patient on physical examination reveals most often tenderness and muscle guarding in the right lower quadrant. Occasionally, the discovery of a soft tender mass or a thickened loop of bowel in this area raises the question of appendiceal abscess. The possibility of appendicitis usually leads to a fairly prompt exploration at which time the characteristic findings of Crohn's of the ileum or ileocecal area are discovered by accident, after which a more detailed clinical history often discloses a history of some longer periods of intermittent recurrent abdominal pain and cramps and a more prolonged history of disturbances of bowel pattern including bouts of diarrhea.

Another classic mode of presentation is the younger patient with a history of abdominal discomfort and periods of diarrhea between episodes of constipation, in the absence of a thorough work up, which usually means frequently not having had a small bowel series, who has been carried for some years as a typical example of the irritable bowel syndrome. In this individual the onset of a complication: (high) fever, right lower quadrant signs of peritonitis, the development of a perineal abscess and fistula, marked rectal bleeding or the appearance of extraintestinal manifestations such as arthritis or pyoderma, raises for the first time, the suspicion that he or she may indeed be harboring Crohn's disease, and so the necessary subsequent radiographic or endoscopic study confirms or establishes the diagnosis.

A third well-recognized entity is the young student who fails to grow or mature sexually or has the delayed onset of menstruation. While anorexia and weight loss may be in the forefront of the clinical picture, disturbances in bowel habits, even abdominal pain, may be absent, and interest in the intestinal origin of this syndrome may not even be raised. If to this is added low-grade fever, some joint pains, and the hemic murmur of anemia, the label of rheumatic fever seems most appropriate. Or the youngster may be noted to have a stiff spine and thus be considered to have juvenile rheumatoid arthritis (Still's disease) without the physician looking for a possible association with Crohn's disease. I must admit that this error is seen much less frequently in my experience in recent years.

A quite characteristic clinical presentation of Crohn's disease is that of a young person who has had the acute onset of a perirectal abscess, treated by incision and drainage which either develops into a persistent abscess or fistula, or the wound fails to heal. Usually investigation of the remainder of the bowel has not been done since the acute rectal pain has precluded this type of investigation. Often the patient has had several attempts at local treatment of the perirectal wound, the first usually in the surgeon's office with inadequate drainage. The chronic indolent course which fails to heal then leads to the suspicion that the patient may have Crohn's disease.

Free Perforation

On rare occurrences a free perforation of the large bowel in Crohn's disease may occur. The tendency of loops of inflamed bowel to become agglutinated and the productive desmoplastic reaction that takes place in Crohn's disease renders free perforation less likely than in the thin, attenuated bowel of ulcerative colitis. The perforation may take place through an area of grossly normal mucosa adjacent to a stenotic area of ileitis or Crohn's colitis. Perforation as the first manifestation of Crohn's disease must be extremely unlikely.

Free perforation in ulcerative colitis is almost invariably a result of toxic colitis and dilation. In Crohn's disease, free perforation or perforation of an abscess usually does not occur as the consequence of dilation. In about a thousand patients with Crohn's disease seen over a 20-year period at The Mount Sinai Hospital, there were about 21 perforations (15 were free perforations of small or large bowel, and six were primary perforated abscesses). Treated by resection with or without any diversionary surgery, they all survived. The clinical picture of sudden severe abdominal pain, disappearance of all small bowel sounds, abdominal rigidity, the demonstration of air under the diaphragm with loss of liver dullness, is characteristic and usually easy to recognize. Sometimes steroid therapy makes recognition of the free perforation more difficult. What role intestinal obstruction or stricture formation plays in these perforations is unclear from our own material, but these factors must play a role in some instances, infrequent, especially in those in whom the perforation did not occur in the area of most active disease. But in a disease like Crohn's where fibrosis is the important consequence of inflammation, perforation is on the whole a rare complication with fistula formation—the extension of the disease into adjacent and agglutinated loops—the rule.

Crohn's colitis occurring as *toxic dilation* of the colon is discussed in the joint section on toxic dilation and megacolon.

Atypical Presentations

It seems that we learn more from and remember better the atypical presentation, although most of the Crohn's patients I have seen have had the more typical forms. A few of these more atypical cases, however, may be worth recording.

Patients with *recurrent lower intestinal bleeding*, without any systemic or local symptoms, and with repeatedly negative x-ray results, endoscopy of the colon, and labeled red cell scans, have been seen on occasion in my practice at least three times. Laparotomy made the diagnosis on these three and cured this clinical manifestation, although if one had remembered that the mother of the patient had Crohn's of the colon, this might have led to an earlier diagnosis of Crohn's ileitis.

Recurrent labial infections ascribed to Bartholin gland abscesses for some years may be the sole manifestation of Crohn's disease in some young women. A low-grade *persistent umbilical infection*, which may ultimately demonstrate its fistulous nature, has been seen a few times.

Hypertrophy of the gums with or without swelling of the lips or cheeks may be the herald of hitherto quiescent Crohn's disease of the bowel. Biopsy specimens of these oral lesions usually shows the finding of granuloma systems.

A *tuboovarian mass* due to Crohn's disease extending from relatively quiescent ileitis proved deceptive on one occasion. On a few occasions, the patient's known Crohn's disease was so quiescent as to be forgotten by the patient or neglected by the patient's referring physician at the time *pyoderma* was active.

On one occasion, an astute hematologist referred a patient to me because of *marked thrombocytosis* with minimal bowel symptoms, which led to the diagnosis of localized Crohn's of the ileum.

Crohn's disease of the intestine may on occasion not be terribly florid, and the disease manifests itself with a *metastatic presence* in the inframammary or groin area, but this is most unusual.

Silent Crohn's Disease

Every now and again in studying an adult for other reasons we discover the typical x-ray evidence of areas of Crohn's disease, especially in the ileum or ileocecal area and then in retrospect excavate from the patient's memory a history of symptoms often occurring in adolescence compatible with the x-ray findings, which now look quite static.

It ought not surprise us that patients either walk around with undiagnosed Crohn's disease of low activity or bear the evidence of distant episodes. Experience during World War II in the U.S. Army camps revealed the presence of Crohn's disease of the ileum in recruits who were well enough apparently to be inducted into service and yet, as was found on autopsy after traumatic deaths, to have had the disease.

Atypical Intestinal Obstruction

Typically, the intestinal obstruction in Crohn's, slow in onset, rarely becomes complete. Even more rarely does it appear out of the blue, so to speak. The commonest factor that transfers an incomplete stenotic area of Crohn's into a full-blown picture of complete obstruction is the presence of an additional inflammatory feature, an abscess or a fistula. However, curiously, obstructive syndromes have occurred, often on Memorial Day or Labor Day, spring and fall holidays, when patients go on picnics and eat lots of corn on the cob. On another occasion, a patient swallowed a peach pit. On another, a mushroom stalk. Most bizarre was the young student who downed a whole jar of "colossal, jumbo-sized," pimento-stuffed olives at one sitting. While the other food obstructions subsided under conservative management and the patient spontaneously passed the obstructing material, the olives had to be fished out at laparotomy. Occasionally, the typical manifestations of Crohn's disease in the large or small bowel with mechanical obstruction due to stricture or stenotic scarring may be seen in the patient with no obvious recent clinical history to suggest an ongoing inflammation in the many years before the current clinical presentation. In a few middle-aged individuals, a probing history may reveal episodes of diarrhea during adolescence usually diagnosed at that time as part of the irritable bowel syndrome.

More often, in some young teenagers and young adults in their twenties, the first clinical presentation is also one of complete mechanical obstructive symptoms, but with no antecedent history at all suggesting inflammatory bowel disease. Yet their films reveal significant stenotic strictures, especially in the terminal ileum and the laboratories reveal little indices of inflammatory activity. These individuals' case histories raise interesting questions as to the duration of the disease and the length of its "incubation period."

Sepsis

On a few occasions, patients with known stable Crohn's disease have gone into vascular collapse with gram-negative sepsis without any clear-cut intra-abdominal precipitating incident. Blood cultures have grown out organisms derived from the gastrointestinal tract, and the patients have recovered from the shocklike picture with wide-spectrum antibiotics and steroid therapy. We must postulate that a micropuncture or microabscess of the intestine as the portal of entry, although obvious sources could not be detected by conventional methods (scanning or radiographic).

Massive Hematuria

I discuss many of the renal complications of IBD elsewhere in the chapter on extraintestinal manifestation of Crohn's disease (Chapter 28). Here I want to call attention to the clinical features of patients with Crohn's disease who have *massive hematuria*. The announcement of an ileovesicle fistula by urinary symptoms of frequency and urgency, or fecaluria or pneumaturia are well known and recorded. What is not so well known and can indeed be alarming to both patients and physicians alike, is the sudden onset of massive hematuria with the painful passage of bloody clots, difficulty in urination and even life-threatening exsanguination. This may occur as the first manifestation of an ileal–bladder fistula, but not usually. Bleeding occurs in my experience more frequently in the course of a known bladder fistula (usually well tolerated by the patient for a long period of time), and I have usually been able to get by without any direct surgical attack on the bleeding point by having a urologist instill a bladder-irrigation system that removes clots and thus prevents obstruction and its attendant pain. But on one occasion my urologic consultant with much trepidation fulgurated (transuretherally) a moderately sized blood vessel that was pumping away; fortunately, this controlled an alarming situation.

Jejunitis and Ileojejunitis

Involvement of the proximal portion of the small bowel with Crohn's disease presents an intriguing variety of this protean disease. The presence of a lesion solely in the jejunum does and may occur in the absence of lesion of the terminal ileum or the remainder of the gut, but I have never seen this in the duodenum alone, although undoubtedly, it must occur. Lesions of the proximal ileum sparing the distal segment have, in my experience, rarely turned out to be Crohn's disease; most often they are lymphomas. Occasionally carcinoid tumor, retractile mesententis, and even an endometrial implant may simulate Crohn's disease of the mid-ileum alone.

These isolated lesions in the jejunum often with several skip lesions may be present for long periods of time with only the vaguest symptoms and in my experience, often beginning in early childhood. At times, these skip lesions in the jejunum are accompanied by similar lesions in the remainder of the small bowel. After many years, these segmental lesions become contracted and scarred down with fibrotic strictures and dilated loops of small bowel between them, the so-called saddle bags. I am aware of several youngsters explored early in the course of this variant whom experienced surgeons have pronounced inoperable because of the extent of these skip areas, only to be reexplored years later and the disease found to have resulted in a reasonable number of localized, contracted stenotic segments capable of being resected or treated by stricturoplasty. These dilated loops of small bowel in between stenotic segments act as blind loops or intestinal diverticula and result in clinically significant malabsorption syndromes. Treatment of this aspect with antibiotics directed against intestinal bacterial overgrowth has resulted in surprising improvement in the patient's weight, reduction in the foul diarrhea, and vague malaise.

These dilated loops of jejunum are vulnerable to perforation just proximal to the stenotic area, and we have seen half a dozen perforated in our series (the mortality of this free perforation in Crohn's jejunitis is much lower than that due to perforation in the colon). Accordingly we ought to consider treating stenotic areas earlier on than we often do. In many cases, because of the number of strictures, the surgeon may have to leave some behind and untreated at exploration, and the patient should be prepared for this outcome in advance. The use of stricturoplasty advocated by John Alexander-Williams and Emmanuel Lee is superior to local resection in this group of patients, except in the segment of the terminal ileum where most surgeons prefer local resection and ileocolic anastomosis.

Gastroduodenal Crohn's Disease

In my experience, gastroduodenal Crohn's is almost always accompanied by Crohn's disease elsewhere in the small intestine and colon. Patients with this pathologic finding have several sets of complaints. The most frequent is epigastric pain that simulates peptic ulcer pain and is often ascribed to an ulcer, especially in patients receiving oral or parenteral steroids. X-ray and endoscopic examinations do not always make it easy to distinguish Crohn's disease or duodenal ulcerations and erosions. The other symptom is that of gastroduodenal and pyloric channel obstruction with easy satiety, a sense of fullness and nausea, later followed by vomiting and a succession splash on physical examination. Barium films reveal a distended stomach which is not emptying appropriately. Patients with early Crohn's disease of the stomach may have mild, not easily characterized, symptoms of dyspepsia confirmed only by radiographic and endoscopic findings of apthous ulcers of the mucosa or superficial linear ulcerations. I have never seen Crohn's disease of the esophagus, although a few cases have been reported in the literature, but this is rare. I have been told about two patients with Crohn's ulcerative lesions of the mid-esophagus who have developed sinus tracks and fistulas into the tracheobronchial tree. I have always suspected these cases to be tuberculous nodes eroding into the mediastinum and lung. If noncaseating granulomas are found, sarcoidosis of mediastinal nodes as well as Crohn's disease of the esophagus, may need to be considered.

The Appendix and Inflammatory Bowel Disease

Not much is written about the appendix in inflammatory bowel disease probably because there is not much to write about, but the subject is interesting.

The clinical situation in which a young patient with right lower quadrant pain, fever, and leukocytosis is explored with the preoperative diagnosis of acute appendicitis only to be discovered to have regional ileitis at operation is an old story by now. Although it is true that more careful history-taking often gives a needed clue to the correct diagnosis in the form of disturbances in bowel habits preceding the present illness for an indefinite period of time, this is infrequently not the case. I go along with the old adage "when in doubt, take it out," meaning the appendix of course. This can save a great deal of worry in the management of the patient's future course, however, the whole question of the involvement of the appendix in IBD is an intriguing one. There is a feeling generally held by physicians that appendicitis rarely occurs in patients with ulcerative colitis or Crohn's disease, especially not in Crohn's. Indeed, this feeling is so widespread that individual reports of classic appendicitis occurring in the course of known Crohn's disease continued to be published sporadically, along with this kind of experience. In my own experience, I have never seen a case of acute appendicitis in a patient known to me to have Crohn's disease. This impression must have some validity. However, Crohn's disease of the appendix alone does occur presenting with the usual symptoms and signs of acute or subacute appendicitis in the patient explored, the appendix is usually removed and the wound heals well. It is only when the pathologist reports Crohn's of the appendix that this entity is considered. Films of the gastrointestinal tract and sigmoidoscopy with biopsy, colonoscopy with biopsy are hastily ordered, and no evidence of Crohn's disease elsewhere is observed. The stool in the appendix does not contain parasites of the pinworm family, but it does need to be considered. In my few patients in whom only the appendix was involved by noncaseating granuloma systems, the subsequent course did not manifest the typical life history of Crohn's disease. What is more intriguing is that even in red-hot universal ulcerative colitis with active disease from cecum to rectum, the appendix is rarely involved in the process. The appendix of the resected colon is often not even examined and, even if it has been, in my experience, there is little to be seen. The same holds true for the appendix of Crohn's ileocolitis, although we can on occasion fortuitously see x-ray evidence of involvement of the appendix on the barium enema. This, of course, is most rare. We cannot help but wonder if the sheltered environment of this group of lymphoid tissues of the appendix in some way protects it. The benign subsequent course of patients with Crohn's appendicitis diagnosed by a histopathologist's finding of noncaseating granulomas raises the question whether they were indeed Crohn's disease at all, especially since *Yersinia entero colitica* has been reported as inducing granulomas as well. Perhaps these patients were subject to the acute ileitis caused by *Yersinia* infection.

The increasing use of sonography should help in deciding whether or not the patient with universal ulcerative colitis or Crohn's ileocolitis has an acute inflammatory process in the appendix and even makes the diagnosis of Crohn's appendicitis in an individual with known Crohn's disease. A recent personal experience is germane. This young adult of 32 years of age, male, who had a quiescent, localized proctitis, inactive for many years, developed "a unique pain on my right side." He had no fever, no elevated white blood count, and no masses or peritoneal signs could be elicited in the abdomen. Sonography showed the typical target, ring-shaped, inflamed appendix so thickened and swollen that laparoscopic appendectomy could not be done and Crohn's was suspected and confirmed by a classical abdominal appendectomy and the organ examined pathologically. No evidence of Crohn's disease elsewhere was present at laparotomy, by subsequent small bowel x-ray, or by colonoscopy. It is expected that he will fit the picture of Crohn's disease of the appendix and have no further problems.

An Alternative Concept of the Clinical Variety of Crohn's Disease: Its Two Forms

Given the astonishing multiplicity of ways Crohn's disease may present itself clinically, can we categorize these in a more organized and simplified scheme that could unify them and make more coherent their protean manifestations? The idea that Crohn's disease may have two forms is an intriguing one and has an attractive persuasiveness. This concept was stimulated in our institution by David Sacher and findings supported by other groups, including recent Japanese workers, and dating back to work by deDomball suggest that Crohn's disease may present itself either as a "perforating" form or alternatively as a "nonperforating" form dependent on its inherent aggressiveness. The more aggressive perforating form presents itself with sinus tracks, penetrating fistulas to adjacent organs and skin, localized perforations, and localized abscess formation and sepsis. The nonperforating forms are more indolent clinically with the slow development of progressive stenotic scarring of the bowel and the slower development of intestinal and colonic obstruction.

The subsequent clinical course of these two modes are also different. The "penetrating," perforating, or acute variety seems to come to operation earlier than the more indolent, slower obstructing variety and the former develops postoperative reoccurrences more quickly.

My continuing experience is consistent with this kind of classification: patients operated on for well-established stenosis and stricture appear to do well for long periods postoperatively and often come to reoperation only after 10 to 15 years.

In this context, it is interesting that the Rome '91 Working Team classification of clinical subgroups of Crohn's disease suggested three: primarily inflammatory, primarily fistulizing, and primarily fibrostenotic.

III

DIAGNOSTIC MODALITIES

6

Notes on the Physical Examination

The physical examination of the patient seems such a primitive way of obtaining the information needed for assessing the state of inflammatory bowel disease or for suspected IBD. If we review our experiences candidly, what can we learn from the physical examination?

General inspection of the appearance of the patient is certainly the heart of the physical examination—the almost automatic assessment of just how sick the patient before us is. We unconsciously assess the degree of wasting and weight loss, the ease with which the patient enters the consultation room, or the degree of abdominal distention under the bedclothes.

In the head, the presence of conjunctivitis usually in one eye of the nodular localized variety may be a tip-off for IBD. For evidence of iritis, other then photophobia, the gastroenterologist must rely on the slit lamp examination of the ophthalmologist. The presence of a crop of aphthous ulcers (cancer sores) is often the herald of an episode of inflammatory activity of the bowel. Crohn's disease of the gums, in the absence of ulceration, may merely present a hypertrophy of the tissue or may be the very rare, diffuse swelling of the lips.

As we look at the *extremities,* clubbing of the fingers and curvature of the nails in the presence of IBD has always meant to me the presence of significant small bowel disease and only rarely the hepatic complications of cirrhosis or sclerosing cholangitis. Swelling of one lower extremity alerts us to the ominous complications of iliofemoral thrombosis and bilateral pedal edema to hypoalbuminemia. Gangrene or just duskiness of fingers or toes raises the question of vasculitis or cryoglobulinemia and/or cryofibrinogenemia. The patient who complains of joint pains may have only slight swelling of peripheral joints or tender edematous joints with fluid in the knees especially. The stiff back and limited mobility of the spine point to sacroiliitis and/or ankylosing spondylitis.

The *skin* can reveal the erythema nodosum of the shins, pyoderma gangrenosum of the lower extremities, rarely of the upper and occasionally the anterior chest wall, and "metastatic" Crohn's disease in the groin of men and beneath the breasts in women. I routinely look for psoriasis of the skin, especially the elbows, and for psoriatic nails, especially in patients with joint complaints. Cutaneous enteric fistulas conclusively point to Crohn's disease and particularly in antecedent abdominal wall scars of prior operations. Ocasionally fistulas in the umbilicus are secondary to migration along the ligament of Terres.

Careful examination of the *perianal and perirectal area* in children and adults can be rewarding, revealing the site of obvious perirectal fistulas and the surgical evidence of previous attacks of perineal abscesses or fistulas. A thick, juicy perianal skin tag clearly raises the question of Crohn's disease; some at times are large enough to deserve the term "elephant ears."

Although we go directly to the *abdominal part of the physical examination,* this may reveal nothing or everything. Swelling and tenderness localized to a previous abdominal scar or the umbilicus are important clues to underlying inflammation. It is often difficult to distinguish signs of peritoneal irritation from the elicitation of direct physical tenderness of the serosa of inflamed bowel. A localized mass, especially in the right lower quadrant, may represent an abscess, but it is hard to separate from a thickened loop of bowel with adherent swollen mesenteric nodes. A markedly distended colon with the ominous absence of bowel sounds, and the diffuse or localized tenderness of *toxic dilation,* comprise a classic syndrome highly suggestive of a localized perforation.

Obliteration of liver dullness on percussion should be sought especially in patients being treated with corticosteroids on each examination of a patient with IBD. The presence of marked tenderness on rectal digital examination, and blood on the examining finger, point to rectal involvement of both Crohn's disease and ulcerative colitis. While rectal strictures can occur in both diseases, I found them more frequent in Crohn's disease, active or quiescent. Gastorenterologists who listen to the abdomen in the course of their physical examination can hear a percussion splash anywhere in the abdomen in the presence of partial obstruction not merely of the stomach, but also of the small intestine in Crohn's disease. Gastroenterologists may also listen to the heart for the murmur of aortic stenosis, which I feel is so often associated with AV malformation in the gut; or evidence of mitral valve prolapse said to be associated with intestinal motility disorders.

A less distended abdomen may allow recognition of the *fatty liver* or the *slightly enlarged liver* of pericholangitis or early sclerosing cholangitis; *splenomegaly* alerts us to the presence of a concomitant hemolytic anemia.

Examination of the chest, even in the presence of a dry cough of the alveolitis of sulfasalazine, usually reveals very little evidence of pathology. The transient pericardial rub may be a reaction to sulfasalazine and may also represent the manifestations of vasculitis.

These are the kinds of findings we may pick up while examining the patient with inflammatory bowel disease of either variety. They are interesting and alert us to the various possibilities, but they need the augmenting feature of more probing techniques, as well as related laboratory studies, to allow us to make the correct diagnostic assignment and interpret the degree of impairment which the patient suffers in an attempt to forecast his or her outcome.

7

The Essential Investigations

I think all would agree that the sigmoidoscopic examination is an essential part of the study of every patient suspected of inflammatory bowel disease and not only because it ought to be part of any complete physical examination. In my practice, at the first interview and examination, I attempt to visualize the rectum and sigmoid colon at all possible without prior cleansing. This often does not allow for a deep penetration, but does permit the visualization of some rectal and colonic mucosa without the trauma of cleansing enemas and cathartics. This examination confirms the diagnosis of ulcerative colitis when limited to the proctoscopic form or the more extensive, left-sided and universal forms.

For Crohn's disease involving the colon, better visualization is required in my experience if the early lesions are to be recognized, especially the shallow aphthous ulcerations. We learn to distinguish these forms of IBD only by repeated experience using standard atlases of reproduction in color of which many exist but differentiation may be difficult. I am a firm believer in doing mucosal biopsies if endoscopic procedures are at all indicated providing the investigator is prepared to handle possible bleeding. Equally important is the necessity of having an experienced pathologist available to examine the tissues. Perhaps even more important is that he or she should be interested in this area of pathology, especially if the question of collagenous or microscopic colitis is at issue.

The Essential Investigations

General Laboratory Studies

In the initial assessment of the patient suspected of IBD, I content myself with a complete blood count, including platelets, serum albumin, sedimentation rate and/or C-reactive protein, and stools for occult blood if there is no history of rectal bleeding, as in Crohn's disease, or if no blood has been detected on the rectal digital examination or following the attempts at signoidoscopy. These are my minima at this stage of the disease and in the planning of therapy.

The Search for Specific Forms of IBD

I have heard it argued that each case of a patient with a new history of a diarrheal disease, with or without an origin of fairly recent onset, which raises the question of inflammatory bowel disease, must be considered as an example of infectious diarrhea until proven otherwise. In general I share this view, but the question always arises as to how far the physician should go in his effort to rule in or out a specific viral, bacterial, or parasitic etiology. This remains a difficult question and I believe we must maintain a flexible attitude in regard to this point.

At the initial presentation, if the patient has fever and is systematically sick, or has returned from foreign travels to endemic areas of specific forms of enteritis and colitis, or has been exposed to antibiotics, or relates the diarrhea to a specific meal, or has had more than casual contacts with individuals with the same complaints, I have the stools examined for *Salmonella, Shigella, Yer-*

sinia, Campylobacter, and *C. difficile* toxin and looked at carefully for parasites, especially amoebic. The difficulty is to find reliable laboratories, but they do exist.

With evidence of amoebiasis, the current serological studies give me some help and comfort, but, in the end, histological examination of rectal or colonic mucosal biopsy specimens properly stained can often settle the question of amoebic infection or concomitant amoeba in the presence of nonspecific IBD.

I do not routinely smear for gonococci, culture or look for cytomegalic virus, look for serological evidence of hepatitis B, herpes virus, or chlamydia, but biopsy may reveal viral inclusion biopsies or atypical mycobacteria of the avium (intracellular) type. Perhaps I should. However, in homosexual and bisexual patients, these searches are a must, as well as the search for giardiasis along with amoebiasis.

Some Observations on Imaging Techniques

In choosing among the variants of imaging techniques in our first effort to establish the diagnosis of IBD, one problem may be an embarrassment of riches. We have a great many to choose from: radiography, endoscopy (flexible sigmoidoscopy or colonoscopy), sonography, computerized tomographic scan (CT scan), labeled red cell scan, angiography, and variants of the labeled leukocyte scans. All are useful. Their application is the issue.

Radiographic Studies

Receiving my clinical experience in IBD before the endoscopic era and with the benefit of having available an outstanding gastrointestinal radiologist, Dr. Richard Marshak, for many years, my bias naturally favored the use of radiographic visualization of the gastrointestinal tract. My recent experience, together with exposure to a world-class endoscopist, Dr. Jerome Waye, has clearly tended to correct my original biases. It would be too schematic to formulate my current view at present as radiology of the small intestine, endoscopy for the colon, but that might be an approximation.

In patients with acute illnesses suspected of having acute ulcerative colitis, I see no need for a barium enema. Its performance may be dangerous and precipitate toxic dilation. I worry too about the use of harsh cathartics in the preparation for a barium enema in a patient with stable ulcerative colitis. With patients in more chronic phases of ulcerative colitis, since they are usually young, I try to avoid an overexposure to radiation in their expected long course and wonder how much radiation may have contributed to the risk of subsequent cancer of the colon in others. I have reported five cases of acute promyelocytic leukemia in patients with IBD whose exposure to radiation was rather large.

There is a place for radiographic investigation of the colon in instances where endoscopy has recorded a mechanical stricture of the colon even if that stricture can be bypassed with the pediatric endoscope. Further, in the attempt to distinguish between adenomatous polyps and inflammatory polyps, a barium enema may contribute useful information. The contour defect, as Marshak preached, indicates a neoplasm. For Crohn's disease of the colon, suspected or documented, I favor endoscopy as a method of choice, but there is a place for the barium enema, especially in the search for fistulization.

In every case suspected of Crohn's disease, I believe a small bowel series is needed and required to establish the diagnosis. In patients with endoscopically demonstrated Crohn's colitis, small bowel films are indicated to document the presence or absence of ileal involvement, the finding of which will be of great importance in future management. Although it should be pointed out that in some instances follow through films of the colon from above are at times more revealing than the barium enema with its overdistention and obliteration of the mucosal pattern.

The Role of Endoscopy: Confessions of a Nonendoscopist

This section is aptly called "the confessions of a nonendoscopist." As a gastroenterologist, I consider sigmoidoscopy as part of the general physical examination. Beyond that, I do not perform any endoscopic procedures.

ENDOSCOPY IN ULCERATIVE COLITIS

In the earliest stages, whether in mild or more severe forms of ulcerative colitis, sigmoidoscopy or even proctoscopy will make the diagnosis often during the patient's first office visit and even without preparation. If there is any question about the findings, repetition after gentle enemas can be done taking care to evaluate properly the effects of these cleansing maneuvers. When the physician is concerned about the presence of amoeba and direct smears from stool examination have not answered the question definitively, rectal biopsies may help. In the milder setting I can see a place for flexible sigmoidoscopy, which can add a little further visualization and define the geographic extent of the disease beyond the rectum or rectosigmoid. However, a positive sigmoidoscopy or barium enema will satisfy my sense of diagnostic rigor in this context.

I have not found rectal biopsy to help me in determining if the acutely inflamed rectosigmoid is that of the onset of ulcerative colitis or simply the beginning of an acute self-limited colitis whose etiology I would suspect is infectious, but is not often confirmed, despite the claims of some pathologists to be able to differentiate them easily.

Paradoxically, if the clinical features suggest that the colonic disease involves the left side of the colon or even the entire colon, I see no need for colonoscopy or flexible sigmoidoscopy and fear the complications of perforation, however low the statistical incidence may be. However, I would like to have most of my patients with ulcerative proctitis have a flexible sigmoidoscopy at some time in their course to determine the exact extent of their localized disease. The information may be quite important if they worsen clinically and may be more important and play a role in determining the question of their long-term endoscopic requirements for cancer surveillance. In sicker patients, especially ones sick enough to be treated with steroids or hospitalized especially for parenteral steroids, as I have indicated, I see no place for more extensive endoscopic examinations in ulcerative colitis and fear the possibility of perforation of the attenuated inflamed mucosas. Certainly, any evidence of dilatation of the bowel or a stretch of continuous air column on flat plate contraindicates any of the endoscopic procedures, including sigmoidoscopy. The continued management of a patient with ulcerative colitis does not require repetition of proctoscopy and/or sigmoidoscopy to evaluate the results of treatment.

If rectal bleeding or the clinical history raises the question of IBD and the sigmoidoscopy and the barium enema have not disclosed any clear sources of rectal bleeding or defined pathology, colonoscopy to the cecum is in order (a careful, thorough search for the possible local anal and rectal sources of bleeding or disease is presumed at this point). The discovery of a patch of erythematous inflamed mucosa beyond the reach of the sigmoidoscope by this approach should lead to a biopsy of the area. Biopsy of the right side of the colon or rectum can also reveal the relatively rare individual with microscopic or collagenous colitis.

Discovery of a polyp or mass lesion in the course of an initial work-up clearly demands the appropriate endoscopic and biopsy procedures to determine the possible neoplastic nature of the lesion, as well as the study of the entire remainder of that patient's colon.

The role of endoscopy in the surveillance of patients with ulcerative colitis for cancer is becoming clearer with increasing experience. The risks of the population to be watched have been better defined. The earliest cancer I have seen in associated ulcerative colitis occurred after a seven-year history of symptoms. In addition to universal disease, left-sided disease requires watching for these patients do develop cancer, possibly at a slower rate than required for universal disease on the average. The pathologic criteria for separating the dysplasia of neoplastic significance from that due to inflammatory activity have now been well defined by a consensus of pathologists. The current timing of repeated colonoscopy and multiple biopsies is rational, but it has not been shown to be based on firm scientific evidence. Indeed, the use of colonoscopy annually for this purpose is based on inferences derived from the presumed doubling rate of cancer of the colon, which, in turn, is based on the doubling time of missed cancers or pulmonary metastases. No one knows what the doubling time of cancer in ulcerative colitis is, but annual observation seems reasonable.

ENDOSCOPY AND CROHN'S DISEASE

In the classic case of Crohn's disease of the small bowel, the barium meal makes a diagnosis. I am not convinced that small bowel enteroclysis improves the chances of making the correct diagnosis missed on conventional small bowel series. On rare occasions, upper intestinal endoscopy performed with the newer push enteroscope does allow a glimpse of the upper jejunum and has revealed aphthous lesions not seen on x-ray. The biopsy specimens have even on rare instances yielded granuloma. On some occasions, it is worthwhile to reach the ileum with the colonoscope to answer the question of the nature of an isolated ileal lesion whose character is not apparent on diagnostic films.

In the typical case of Crohn's disease of the anus, rectum, and sigmoid, local endoscopy makes the diagnosis easily and biopsy may help decide the issue. When a barium enema reveals the classic finding in Crohn's disease of the rectum or sigmoid, endoscopy does not, in my opinion, add appreciably to the diagnostic yield. Colonoscopy has been useful in answering the question of whether the cecum is involved with clear-cut Crohn's disease of the ileum, especially a point to be noted if surgery is contemplated. If the ileal disease is somewhat atypical, finding a granuloma on the rectal biopsy specimen in 10 percent of patients with Crohn's disease whose colons are normal on endoscopy or x-ray examination does strengthen the clinician's hand. Periodic colonic endoscopy, partial or entire, is valuable for following the response of therapeutic measures and spares the patient the exposure to radiation, which undoubtedly will continue in his or her long life with Crohn's disease. The exact role of endoscopy in the surveillance of patients with Crohn's disease for cancer is yet to be determined, but I believe it will be important as the dysplasia of Crohn's colitis is better defined and recognized.

Scans: Structural and Functional

The fundamental techniques for diagnostic imaging remain radiographic and endoscopic, but the scanning techniques do have their place in the evaluation of complications of Crohn's disease.

Radionuclear scanning has two areas of interest for me. First, the *radionuclear colloid scan* of the liver has a limited role in the search for liver abscess or multiple abscesses as a source of fever in patients with IBD now superseded by the CT scan. In my experience, the diagnosis of liver abscess, especially in Crohn's disease, is often made late in the hospital setting and is not thought about frequently enough. Liver abscess in ileocolitis is not rare and can be seen at any time in the course of the disease. The difficult problem is the insidious onset of painless fever in the post-operative state. Liver scan has its valuable place in combination with the lung scan to determine the presence of a subphrenic abscess of the right side. Second, *the labeled red cell scan* for visualization of gastrointestinal bleeding in IBD has a limited role. In ulcerative colitis, bleeding is diffuse and rarely localized. In very few instances, this technique was of help in deciding which of several skip lesions in Crohn's disease was the source of bleeding. Ultrasound is useful in locating masses, but has a very limited value in distinguishing between matted loops of bowel within the abdomen and swelling from abscesses.

CT Scanning

Increasing access in the use of the CT scan has convinced me of its importance in the management of IBD, especially its complications.

In the presence of equivocal x-ray films, especially of the small bowel, CT scan has increased the likelihood that the lesion of the small bowel wall is due to Crohn's disease. More important, CT scan has proved invaluable in distinguishing an abscess from a localized inflammatory mass in Crohn's ileocolitis or colitis. More recently I have been less reluctant to request needle aspiration of documented intra-abdominal abscesses in Crohn's disease because of the fear of inducing fistulas of the skin. Perhaps this is too cautious a stance, but I believe that, given a segment of Crohn's that has led to an abscess, the abscess should be drained and the segment removed en bloc if at all possible. A vexatious area in which CT scan may be felt to play a larger role and perhaps a more helpful role is in our attempt to distinguish strictures of the colon in both Crohn's disease and ulcerative colitis as being benign or neoplastic in origin. More equivocal has been my experience not with strictures but with areas of narrowing in the colon in Crohn's where the CT scan can be helpful in distinguishing a neoplasm from an inflammatory cause, and in perineal fistulas.

Labeled Leukocyte Scanning

Labeled leukocyte scans have been available to me for routine use for a considerable time now. While its use in some instances in attempting to quantitate the degree of inflammation present in Crohn's disease has been helpful, its diagnostic use seems to me rather limited. I have attempted to use it to determine whether other areas of the large and small bowel in patients with Crohn's, but not obviously detected by radiographic or endoscopic evidence, were also the seat of inflammation but without remarkable success. I have found this method useful in some instances where the patient required prompt operation, but when no radiographic or endoscopic study was available.

Diagnosis of Perianal Disease by Imaging Techniques

Perianal and perirectal disease in patients with Crohn's disease elsewhere in the gut present no diagnostic problem when an obvious abscess or perianal fistula or rectovaginal is present. The problem is more difficult when the patient complains of pain and local examination either digitally or endoscopically is painful and often not completely revealing and an examination under anesthesia seems rather excessive. Endosonography, CT scan, and magnetic resonance imaging (MRI) have all been reported as useful with varying claims of superiority for each of these modalities. CT has helped me in very few instances; I have a limited personal experience with the endosonography or MRI, but the latter does have some theoretical advantages of allowing visualization in two different planes and I expect will be used more frequently in the future in obscure situations.

IV

PROBLEMS IN DIAGNOSIS

8
The Limits of Accuracy

In making a diagnosis and assigning the patient to a specific category of inflammatory disease, we must remember that we are giving a tentative label only and must be prepared to reassess this label as further information becomes available. Based on clinical grounds, it is possible that the indeterminate group even for the most experienced physicians may reach up to 15 to 20 percent of cases, but the subsequent course of the natural history of a particular patient's disease will usually make the diagnosis firmer. It is salutary to remember that this indeterminate group may be larger than 10 percent even when the pathologist has adequate tissue material at hand.

Some index of the difficulty in sorting out Crohn's disease from ulcerative colitis is illustrated by my former experience in selecting patients for the Kock continent ileostomy, which I had reserved for ulcerative colitis only. In selecting more than 50 candidates for the operation, three patients who appeared to possess the requisite characteristics of ulcerative colitis (with which diagnosis the knowledgeable surgeons involved concurred) turned out at laparotomy to be more characteristic of Crohn's disease by gross inspection, did not have that operation, and the diagnosis was later confirmed by microscopic examination of the resected material. Two of the entire group considered by both gastroenterologist and surgeon to have ulcerative colitis whose disease were deemed suitable for the continent ileostomy that was performed at operation turned out on microscopic examination to have "classic Crohn's disease."

In the absence of this kind of opportunity to examine the entire colon by a pathologist, it is difficult to know how many subjects remain in the indeterminate group with time. The development of a perianal abscess or fistula, clear-cut radiographic evidence of small bowel disease, microscopic involvement of the ileum as discovered by endoscopic biopsy, all shift the diagnosis from the ulcerative or indeterminate group to the Crohn's category. The development of significant disease in an ileostomy, diagnosed by either retrograde endoscopy or gastrointestinal x-rays, places the patient whose colon could not be easily labeled in my opinion into the Crohn's category. Certainly, the uneventful life of an ileostomy favors the diagnosis of ulcerative colitis, but this is far from being a terribly convincing bit of evidence.

It is part of my dogma regarding the chronic nature of Crohn's disease that complete radiographic clearing of ileal disease makes me reject the original diagnosis and believe that the patient suffered the resolution of an acute appendicitis or of a *Yersinia* infection. Restated, I believe that every episode of acute ileitis in Crohn's disease is part of a chronic process.

The increasing use of the surgical creation of an ileal reservoir pouch for the treatment of ulcerative colitis and the significant incidence of a nonspecific inflammation in this pouch, so-called pouchitis, has made the accuracy of our diagnostic labels of ulcerative colitis as opposed to Crohn's disease more than an academic exercise. Most examples of this inflammatory reaction in the pouch respond to current therapy with antibiotics. Failure of a small group of more refractory instances, so-called refractory pouchitis, is often ascribed at present to an error in diagnosis of the original colitis, a failure, that is, to diagnose Crohn's disease correctly. Our recent review of Mount Sinai Hospital cases does not support this conventional explanation and the results of the blindly controlled re-examinations of the colons are of some interest in this context. The resected colons of 33 patients in whom pouches were created were reclassified pathologically. Of these, five are now considered to have been Crohn's disease and bore no relation to whether or not the subsequent pouch developed significant inflammation. This reclassification offers some indication of the limits of accuracy of preoperative–histologic examination of colonic biopsy material. The even more recent finding, by Mount Sinai investigators and others, of a high prevalence of supposedly ulcerative colitis–specific antineutrophilic cytoplasmic antibody (ANCA) among these refractory pouchitis cases, militates even further against the conventional wisdom that refractory pouchitis is simply a reflection of underlying, misdiagnosed Crohn's disease.

9

Making the Difficult Diagnosis

In the usual variants of *ulcerative colitis,* ranging from proctitis to universal disease, it is ordinarily not difficult to make the diagnosis. Sigmoidoscopy performed in the patient with suspected ulcerative colitis usually makes the diagnosis easily on gross inspection even before the biopsy results are reported from the pathology laboratory. Somewhat more difficult is the patient with disturbances in bowel movements and with rectal bleeding and in whom conventional double-contrast barium enema and sigmoidoscopy, flexible or rigid, are well within normal limits. The history of gross bleeding should prompt colonoscopy to the cecum. In some individuals, occult bleeding with red and white blood cells, especially eosinophils in fecal smears, is also a good indication for a flexible endoscopy of the entire colon. The discovery of a patch of erythematous, somewhat friable mucosa that on biopsy reveals "nonspecific acute and chronic inflammatory changes," reassures me that we indeed can call this disorder a segmental colitis. But is this disorder related to ulcerative colitis? Only further observation of the patient's course will make the diagnosis more secure with time. Some forms are what for a better term we most consider "acute self-limited colitis." Perhaps these are abortive episodes of ulcerative colitis in the same way that acute infectious mononucleosis might be considered as an "aborted" leukemia. Some perhaps are transient self-limited infectious forms whose causative organism has so far eluded us or has not been searched for.

Ulcerative Appendicitis in Ulcerative Colitis

While the clinical incidence of acute appendicitis is generally held to be quite rare in inflammatory bowel disease, not much attention has been paid to the problem of the involvement of the appendix in ulcerative colitis. Based on the older Scandinavian literature, I had the impression that, in ulcerative colitis involving the whole colon, one-third of the specimens of the resected colons in ulcerative colitis showed actual ulcerative disease in the appendix, one-third were normal, and in the remaining third, the appendix was obliterated. My impression needs to be modified in the light of the recent studies done at our institution by the pathologists Groisman and Harpaz, and the gastrointestinal fellow, George. In 160 consecutive adult and pediatric patients with ulcerative colitis whose colons were removed, 45 of these were obliterated. Of the 115 other specimens in both adults and children, ulcerative appendicitis was present in 107 whether the patient had suffered from universal or localized disease. Similar findings have recently been reported in children with both Crohn's disease and ulcerative colitis. This may help explain some instances of right lower quadrant pain in individuals with left-sided ulcerative colitis.

Making the diagnosis in some forms of *Crohn's disease* may be quite difficult. The adolescent or young adult with moderate systemic symptoms clearly sick with some weight loss and altered bowel patterns of a soft or loose stool frequently away from home for the first time as a freshman in college should by all rights have Crohn's disease. But what are we to do in the face of negative films of the entire tract and no aphthous ulcerations on barium enema with a normal rectosigmoidoscopy. Here colonoscopy has been of considerable help. Finding areas of localized tiny ulcerations in the ileum if the colonoscope can get into it or aphthous ulceration helps and their biopsy allows us to make the diagnosis of Crohn's disease. While one granuloma does not make a disease, the discovery of a granuloma system in my opinion makes the diagnosis of Crohn's disease more secure. The 10 percent positive rectal granuloma found in patients with normal gross rectosigmoid mucosa and negative barium enema with known Crohn's disease of the small bowel should lead the physician to do biopsies routinely in situations of suspected but unproven Crohn's disease. Using the colonoscope to reach further down the small intestine from above has allowed my colleagues to see at times scattered ulceration not seen on the small bowel barium studies, including enteroclysis, and at times to produce biopsies that contain acute and chronic inflammatory cells and even rarely granulomas in patients suspected of Crohn's disease but with negative x-rays.

Should a physician ever make a diagnosis of Crohn's disease without radiographic or histologic support? One can certainly suspect the diagnosis, but it is still a hazardous undertaking with the risk of prematurely and erroneously labeling a lifelong illness. Yet I have seen at least two people, one a physician, suspected of having Crohn's disease but never documented who turned up several years later with classic x-rays and pathologically documented Crohn's. I have been informed of one patient in whom the diagnosis of clinically suspected Crohn's of the small bowel resisted all techniques of confirmation until laparoscopy revealed the classic fat creeping over the bowel. A rare bird certainly, but I wonder if this modality could not be used more frequently.

Even more difficult to diagnose is the patient with strong clinical features and only equivocal x-ray of Crohn's disease in whom endoscopy has not really been helpful. Should the physician have the patient explored or given a trial of medical therapy? While most antibiotics seem relatively harmless drugs to try, steroids and immunosuppressant drugs are too hazardous to use for this diagnostic approach. I know of two such patients who were explored. In both, no gross lesions but only slightly enlarged lymph nodes were discovered in the mesenary of the large and small bowel. On biopsy, the findings of granulomas in these two instances raised more questions than answers. Neither of these two individuals went on to develop the more classic findings of Crohn's disease. Concomitant liver biopsy in one was reported negative for granuloma systems as well.

10

Differential Diagnosis and Errors in Diagnosis

Most diagnostic medical errors I believe are errors of omission. What we do is essentially rational. It is what we do not do that is important. The basic mistake we make in diagnoses is not to make our differential diagnosis wide enough. In this connection, the commonest error I have seen is the failure to think about inflammatory bowel disease, especially Crohn's disease, either because the patient is too young (under age 10) or too old (after age 55 or 65) or because the history is atypical, especially if extraintestinal manifestations dominate the picture—for example, the patient with arthritis involving multiple joints or with florid skin manifestations.

The second commonest omission is the failure to obtain a complete small bowel series and then with only an upper GI series at hand to interpret some minor findings in the duodenal bulb as evidence of a duodenal ulcer. Further, if the history is one of abdominal distress and change in bowel habits and evidence of colonic bleeding are present, the physician ought not to dismiss the diagnosis of ulcerative colitis even if a barium enema or rigid sigmoidoscopy reveal no evidence of inflammatory changes. Here I have found colonic endoscopy may reveal a patch of segmental colitis, which I include in the family of ulcerative colitis, if the small bowel is negative, since these patients rarely go on to present the full blown picture of Crohn's disease.

All diseases of the ileum are not Crohn's ileitis. Hodgkin's disease and lymphoma are serious disorders that commonly mimic Crohn's ileitis or ileojejunitis. About every two to four years in our institution, but rarer now, the specimen of ileum or ileum and cecum resected for Crohn's disease turns out to be tuberculosis. Primary cancer of the ileum does exist even in the absence of Crohn's disease. Metastatic cancer to the small bowel, especially from breast as well as from melanomas, can present a confusing radiological picture. A plasmacytoma of the ileum can imitate ileitis quite deceptively. Carcinoid tumors of the ileum may certainly mimic ileitis.

Infiltrative diseases of the small bowel, including amyloid, as well as lymphomas, have been mistaken for ileojejunitis. The nodular lymphoid hyperplasia of the duodenum with giardiasis and dysgamma globulin anemia on occasion can erroneously be diagnosed as Crohn's disease. The lymphoid hyperplastic nodules of the terminal ileum in normal young people are rarely nowadays confused with ileitis but the mistake continues to be made.

A small cancer of the cecum not easily seen on barium studies, even with air contrast, may extend to the ileum producing a rigid stenotic ileum difficult to distinguish from primary inflammatory disease.

We all know that acute ileal changes can occur in infections with *Yersinia* and *Campylobacter* infections. But the error I have seen is to label these young people with a diagnosis of Crohn's because the local laboratory failed to report the presence of the organisms or has not been asked to look for them. Since Crohn's disease carries with it a serious long-term connotation, we must avoid premature labeling, especially in these young subjects.

A quite common cause of ileal radiographic alterations is the presence of a clinical mass or x-ray evidence of one by indentation of the cecum. This, of course, may be due to an acute appendicitis or an appendiceal abscess in which case sonography may be helpful. While the x-ray appearance of the ileum can return toward normality in Crohn's disease on rare occasions, and indeed a few observers have reported it returning to normal, this occurrence is a most unusual course. Therefore, as I have said before, the possibility that the patient is recovering from an appendiceal abscess or *Yersinia* infection must be raised in those instances in which the x-ray appears to become completely normal along with the concomitant "cure" of the patient.

In addition, radiation of the small bowel can lead to a form of chronic enteritis. I have seen this often following x-ray treatment for ovarian neoplasm. More external fixation and angulation of the small bowel than is the case in Crohn's ileitis is present along with minimal systemic symptoms, except for obstructive pain and protein leak. While nowadays great effort is expended to avoid such inadvertent radiation, fixation of a loop of bowel in the exposed therapeutic field may be the result of adhesive disease from previous operations, especially appendectomy or lymph node dissection for other reasons.

The late Leon Ginzburg resected the ileum of a young man of 40 because of obvious stenosis considered preoperatively to be his form of ileitis and looking like "burnt out" ileitis that turned out at the operating table to be an instance of radiation enteritis following x-ray therapy for testicular carcinoma with negative bilateral lymph node dissection some 20 years before while the patient was in the military service. This past history was entirely forgotten by the patient himself.

In addition to such neoplastic processes as lymphoma and Hodgkin's disease already cited above, the nonfunctional carcinoid tumor of the ileum should enter into our differential diagnosis.

Infarction in the small or large bowel can obviously present with rectal bleeding, abdominal pain, and peritoneal signs and be mistaken for Crohn's or ulcerative colitis. Formerly seen quite frequently in young women on the contraceptive pill, this variety fortunately has markedly decreased with the reduction in the estrogen content of the current contraceptives. Sigmoid diverticulitis ought to be easily separated from Crohn's disease or even ulcerative colitis in the average patient with left lower quadrant pain and localized tenderness perhaps with low-grade fever, but may be difficult on occasion. The presence of rectal bleeding in a patient with a relatively spared rectal segment literally raises the question of a localized colitic process, possibly even Crohn's disease. On endoscopy, the presence of a few sigmoid diverticula in the middle-aged patient may be noted and given undue weight in the differential diagnosis. A relatively short or long submucosal tract may not solve the problem of labeling the disorder. Of course, a sigmoid bladder fistula can contribute to further confusion, including the development of some extraintestinal manifestations in diverticular disease, so this differential diagnosis may indeed be difficult to make.

Errors in Ulcerative Colitis

Gastroenterologists tend to assume by the time they get to see the referred patient that inflammatory colonic disease of a particular individual falls in the nonspecific ulcerative colitis group. Errors in this group are often due to failure to search diligently for *Entamoeba histolytica*. Here the physician may need to look carefully at rectal and colonic biopsies for any organism in situ to make the diagnosis. The serological test has not been of much help to me.

Well recognized at present is the error of assuming that a patient with clinical ulcerative colitis has Crohn's disease because the rectum is relatively spared, especially after rectal steroid or 5-ASA installation. I have already considered the need for detailed investigation in that group of patients who have a segmental colitis not seen on routine sigmoidoscopy. In the present era of emphasis on cancer in ulcerative colitis, the error in missing this diagnosis is less likely to be made with the use of more frequent and multiple biopsies for dysplasia. I have the impression that dysplasia is now less overdiagnosed and overutilized in our anxiety to avoid missing a cancer, as the criteria for this marker have been stabilized and the inflammatory type separated out more clearly.

Well aware that a stricture in ulcerative colitis must be considered suspect for cancer, we tend to relax in the presence of a stricture of the patient we have perhaps labeled erroneously as Crohn's disease of the colon. Often the colonic disease may fall into an indeterminate group by the time we see the patient. It is in this situation that a neoplastic stricture may be misinterpreted as benign and we should remember that strictures in Crohn's disease are also malignant at times.

Errors in Patients with Crohn's Disease

Although the terminal ileum may be spared in Crohn's disease of the jejunum or in a few instances of proximal jejunoileitis, the diagnosis of Crohn's disease of the mid-small bowel in the absence of radiological evidence of involvement of the terminal ileum does give me great concern. I have seen the following kinds of errors in patients with radiographic mid-ileal lesions mimicking Crohn's disease:

1. The middle-aged female with biochemical findings suggesting a functional tumor before the appearance of clinical findings or evidence of the carcinoid syndrome.
2. Retractile mesenteritis in a young male whose only systemic symptom was low-grade fever.
3. A mid-ileal lesion obstructing the lumen considered to be Crohn's for several years was treated as such in a physician's wife of childbearing age which on resection was found to be an endometrioma.
4. Several isolated mid-intestinal lymphomas.
5. Several patients with erosive ulcerative disease of the duodenum and jejunum have been referred to me as duodenal and jejunal Crohn's disease with sparing of the remainder of the small bowel and colon whose dilated duodenum and jejunum were characteristic of the *Zollinger–Ellison syndrome* of a pancreatic gastrinoma.

Errors in Patients with Known Crohn's Disease

Patients with previously diagnosed disease present another context for mistakes in diagnosis. One of these mistakes is not to relate the presence of a new finding to the original diagnosis. On several occasions, gynecologists in discovering a pelvic mass have explored these patients only to find that the mass did not represent ovarian or uterine pathology, but was an inflamed segment of small bowel adherent to the pelvis or dropping into the pelvis.

Another striking example of this type of error was in a patient who suddenly developed painful, gangrenous cold fingers and toes. Knowing that this patient had Crohn's disease led to the discovery that this fascinating complication represented concomitant cryoglobulinemia and avoiding angiography as well as the prompt correct institution of anticoagulant therapy.

The more common error is to relate every new clinical feature to the presence of known inflammatory bowel disease. In Crohn's disease, I observed that a carcinoma of the right ovary on several occasions and the right lower quadrant mass that was felt on pelvic examination were considered to be due to the known ileocecal Crohn's disease. The same error was present in a patient where a cecal mass due to his cancer was ascribed also to known inflammatory disease in the ileocecal area. Twice recently, the development of a palpable right lower quadrant mass, in reality a cancer of the ileum, was ascribed to the inflammatory nature of the known underlying disease. The error was to discard the history of an earlier bypass operation.

Failure to extend the colonic flexible endoscopy to the cecum and obtain a biopsy to the right side of the colon has led to the failure to diagnose collagenous colitis and/or "microscopic colitis" in patients usually considered as having irritable bowel syndrome. In one individual in his late seventies, the "irritable bowel" of at least four years' duration was due to lymphocytic microscopic colitis.

Probably the *worst* error is to ascribe a sudden severe total small bowel obstruction to known Crohn's disease and treat it conservatively, when the problem is adhesive obstruction or volvulus. I have seen this once in a young patient with fatal consequences.

Sigmoid diverticulitis at first thought should present little difficulty in the differential diagnosis, yet its distinction from IBD or vice versa can be puzzling at times, especially in the middle-aged patient. Part of the problem arises, of course, from the increasing and widespread incidence of sigmoid diverticulosis in this age group. Further, the confusion of Crohn's disease in the sigmoid is emphasized by rectal sparing, urinary symptoms, including the development of a fistula to the bladder, subserosal sinus tracts, and even the presence of some extraintestinal manifestations of sigmoid diverticulitis; all of these features are, of course, commonplace in Crohn's disease.

If the history includes known small bowel disease and/or perirectal and perianal fistulae, sigmoid diverticulitis cannot easily be discarded from the differential diagnosis. Disturbance of colonic motility, rectal bleeding, fever, localized sigmoid tenderness, even local peroneal signs do not help. Even the gentle limited flexible sigmoidoscopy, which reveals the presence of some diverticular mouths and localized inflammatory changes, grossly and on biopsy may not help also. Even in classic Crohn's disease uncomplicated by the presence of diverticulosis, classic histology, including granuloma systems, is often missing.

So the problem often remains: Are we dealing simply with a patient with localized Crohn's disease of the colon who happens to have some sigmoid diverticula? Our group originally thought that a long subserosal track favors Crohn's and a short one diverticulitis, but this has not been borne out with further observation.

In some instances only the long-term outcome will help to characterize the individual patient as either having Crohn's disease or sigmoid diverticulitis, especially the demonstration of segmental distribution and other areas of inflammation of the colon, periabscess, or ulceration; or small bowel disease will make the diagnosis of Crohn's disease more secure. Inflammatory changes secondary to the antibiotic-associated diarrhea can be confusing, since all patients suspected of diverticulitis receive prompt antibiotic therapy at present. The presence of these possibilities suggest that one should be prudent and somewhat hesitant in the addition of steroids to the therapy of this particular diagnostic crux.

Other Problems in the Diagnosis of Crohn's Disease

I have long held the view that it is hazardous to make the diagnosis of Crohn's disease of the small bowel in the absence of radiographic evidence. The question then arises: What to do if Crohn's disease of the ileum or ileocecal area is discovered at laparotomy for suspected appendicitis in the absence of prior x-rays? All observers agree that the appendix should be removed, but no immediate resection done in this urgent or emergency operation. It is, however, not rare to see patients whose ileitis has not been observed in the small abdominal wall incision for the appendectomy.

In almost all instances, when the radiographs and clinical history are characteristic of Crohn's disease, the gross findings in laparotomy are obvious and typical. It is rare that the surgeon will have any problem at the planned exploratory laparotomy defining the margins of resectable disease in this instance. The question has been raised: What about the rare instances when the bowel looks and feels normal? I believe that if the clinical joint decision before exploration by physician and surgeon has been to resect, the area defined radiographically should be resected. Reverse errors rarely are made. Ileums that grossly look like Crohn's when resected almost always turn out to be Crohn's histologically. The only exception I can recall was a female patient suspected of Crohn's in the era of bypass surgery who, at laparotomy, had tiny nodules resembling tubercles as occasionally seen in Crohn's in whom the biopsy of the small bowel wall revealed Whipple's disease just before the introduction of antibiotics in this disorder.

11

The Obscure Differential Diagnosis

A puzzling differential diagnosis is the occurrence of fresh rectal bleeding in a patient whose history is that of irritable bowel, especially of the spastic and constipated kind. On several occasions I have been puzzled by the finding of *solitary* or *multiple ulcers* in the first 5 to 10 centimeters of rectum. Sigmoidoscopy has usually shown a normal mucosa surrounding the ulcers, and either colonoscopy or films have shown no other source of bleeding. The ulceration is not always present, and as the patient goes from one consultant to another, the ulcerations may or may not be present and may or may not be seen. The solitary ulcer is nondescript in appearance and the biopsy specimens have not really shown anything other than chronic, nonspecific inflammation. Our colleagues abroad refer to this as the *solitary ulcer of the rectum.* In my experience, I always thought this might be due to ulcerations secondary to the hard stool of the rectum or rectosigmoid and I have been persuaded in a number of cases that self-inflicted trauma plays a part. Occasionally an enema-bag tip or a rectal thermometer can do the injury in old or feeble patients. But in younger adults with a history of spastic, constipated colon, the physician must ask carefully and circumspectly about the degree of trauma the patient may have inflicted on him- or herself, especially in those patients who have had to dig out the stool with their fingers. While the solitary ulcer does exist aside from the homosexual world, I think the patient must be cautious against exerting force in digging out stool from the rectum and, with this in mind, rescoping within a month or two should lead to the disappearance of the ulceration. By the time patients get to see me, they have been on sulfasalazine or steroids to no avail. Here I think the treatment of the spastic bowel is a prime consideration and the avoidance of local trauma.

Colonic mucosal prolapse, which must be looked for specifically in the absence of clear-cut rectal prolapse, may also be associated with a solitary rectal ulcer. The pathologist can help us here by finding the presence of a characteristic thickening of a noncontinuous band of collagen, not to be confused with the continuous thickened band of submucosal collagen of collagenous colitis.

I have always considered *collagenous colitis* as a rare disease which probably reflects only my failure to make the diagnosis in the past for I have seen at least five new cases during the last two years. The intractable diarrhea in the otherwise healthy, middle-aged or older patient, whose work-up by the conventional techniques has proven fruitless, should be the telltale clue to look for this entity. Incidentally, one of my patient's literature review indicated that the vast majority of reported patients had been women past the menopause (unfortunately, a trial of self-administered estrogen in this patient was of no avail). The diarrhea in these individuals is not as profuse as in patients with adenoma of the secretory type and is most often osmotic in character. Those I have seen have had extensive work-up and have been treated for idiopathic secretory diarrhea or idiopathic intestinal bacterial overgrowth without much success. The major diagnostic error is not to do biopsies of the right side of the colon when it appeared normal on colonoscopy or to rely on a solitary rectal biopsy when this was done. Although the relationship of collagenous colitis to microscopic colitis is being argued in the literature, my pathology colleagues were not impressed with the degree of inflammatory changes in these cases. Despite the reported successful response from sulfasalazine and the disappearance of the collagen band in a patient ascribed to Mepacrin (atabrine), my patients have not had such good fortune. Trials of sulfasalazine and 5-ASA by mouth were of no avail but oral steroids induced a prompt clinical remission, which was sustained only by continued low-dose steroids, which in one patient whom we restudied did not lead to any resolution of the collagen band.

In this context, it is finally important to remember that patchy, noncontinuous collagen increases in the submucosa of the lamina propria of the colon can occur in both specific and nonspecific inflammatory episodes. One such difficult-to-treat individual had this pathologic confusing finding as the result of an episode of pseudomembranous colitis following antibiotics. I have looked for and found no relation of collagenous colitis to any of the so-called collagen-vascular disorders.

What relation collagenous colitis has to the entity of "collagenous sprue" in which there is villous atrophy of the small bowel associated with deposition of collagen in the mucosa, which is usually refractory to the conventional kinds of celiac disease treatment, is not known, but it is probably not a close one. In one patient with collagenous colitis that I saw, the thickened collagen deposition was present not only in the colon, but in the submucosa of the small intestine. However, there was no villous atrophy and there certainly was no clinical evidence of malabsorption. It has been postulated that the collagen membrane prevents the reabsorption of water from the colon. These patients reinforce my basic feeling that, if a physician performs a colonoscopy looking for a difficult diagnosis, he or she ought to do multiple biopsies even of normal looking mucosa and alert the pathologist to the possibility that they are looking for Crohn's amyloid or collagen deposition, or microscopic or lymphocytic colitis.

Microscopic Colitis

In the differential diagnosis of patients with watery diarrhea in absence of other organic disease and a grossly normal-appearing colonic mucosa by full colonoscopy, I have pointed out the need to separate out from this group of older patients those with collagenous colitis. This entity, with its continuous diffuse collagen submucosal band occurring mainly in postmenopausal females, is difficult to treat despite the published reports.

Since 1980, it has been known that the syndrome of watery diarrhea with a normal colonoscopic appearance may be associated with microscopic inflammation and these individuals have been diagnosed as having "microscopic colitis." Although patients with collagenous colitis have varying degrees of inflammatory cells in the biopsies, the criteria for collagenous colitis have been refined by the work of the Johns Hopkins pathologists and clinicians and can be distinguished from microscopic colitis, although some cases have been reported as progressing from microscopic colitis to collagenous colitis.

The hallmark of microscopic colitis is the histological presence of lymphocytic inflammatory cells diffusely throughout the colon. This diffuse distribution must be distinguished from focal collections of lymphocytic cells which can occur in the presence of polyps or inactive diverticulitis.

Microscopic colitis is not a common disease and the lymphocytic colitic form does not occur frequently. Patients with lymphocytic colitis, in my limited experience, occur in older men and women with no antecedent history of long-standing inflammatory or irritable bowel syndrome. This microscopic appearance may be seen in some patients with celiac disease, but in only one patient with an abnormal small bowel x-ray pattern highly suggestive of sprue with lymphocytic colitis, microscopic criteria did not respond to gluten withdrawal but did respond to 5-ASA by mouth.

It is obvious that the whole area of microscopic colitis, which now includes collagenous colitis, lymphocytic colitis, and focal microscopic colitis, needs further sorting out.

Rare forms of small bowel disease can be confused with Crohn's disease and present puzzling problems in differential diagnosis. In general, the current tendency is to consider most disorders of a localized nature occurring in the distal small bowel as Crohn's ileitis, it seems to me, judging from patients seen in referral. Their failure to respond to conventional ileitis therapy then raises the question of the validity of the original diagnosis. Rarely thought of because it occurs so infrequently is another cause of slowly increasing mid-intestinal mechanical obstruction, that due to a localized stenotic area secondary to the curious inflammatory desmoplastic reaction of the mesentery known as *retractile mesenteritis*. This type of fibrotic mid-ileal obstruction has been seen by me in association with a carcinoid tumor of the nonsecreting type; and in another patient whom I thought the reserpine therapy for his high blood pressure and a well-known serotonin releaser might be playing a role.

These clinical syndromes may be part of a more widespread one of desmoplastic reaction syndromes. Some of these unfortunate individuals may have Riedel's struma of the thyroid or retroperitoneal fibrosis, which may surround the duodenum as well as the mesenteric small bowel. On rare occasions this fibrosis may be associated with a small intestinal carcinoma, one such seen by me in the duodenum of a 13-year-old female.

V

SUMMING UP THE DIAGNOSTIC STANCE AND BEYOND

12

Assessing the Patient's Clinical Status and the Degree of Impairment

Before we can design a therapeutic program for the patient with IBD, which means setting a realistic therapeutic goal of expectations and treatment for the patient and for ourselves, we need to formulate explicitly what we usually do automatically, almost unconsciously. The physician needs to assess the clinical status of the patient in several distinguishable modes—the geography, as it were, of the pathologic involvement in the small and large bowel and its degree of inflammatory activity together with their concomitant physiological impairments, the extraintestinal components, and the impact of the disorder on the patient's total behavior.

But whenever the devices by which we attempt to measure these factors, it is our global impression in the end that we are left with while committees grapple with methods of quantifying the individual components, especially the degree of clinical severity. This is not to deny the value of Truelove and Witts's criteria for assessing severity of illness in ulcerative colitis or the Crohn's Disease Activity Index (CDAI) of the U.S. National Crohn's Disease Controlled Trials which, despite its encountering heavy criticism, remains a useful tool in controlled trials of varied therapeutic regimes, despite its limitation for individual clinical problems and treatment. I do not wish to deny the value also of the simplified criteria of Harvey and Bradshaw. I do not feel that the reliance of the CDAI on several subjective factors is an important defect since the patient's subjective estimates of their degree of illness and well-being are an important component of their presentation to us.

Clinical Assessment in Ulcerative Colitis

I think most experienced clinicians would agree with me that clinical assessment in ulcerative colitis is easier and simpler than in Crohn's disease even when we factor in the extraintestinal manifestations of both.

There is no problem in using the general symptoms of fever, loss of appetite, documented weight loss, abdominal cramps and pain, together with the number of rectal evacuations during waking and sleeping hours. The patient's estimate of degree of bleeding is harder to evaluate, but they can be helped to distinguish the passage of bloody stools from the passage of blood per se or in clotted form.

The physical examination of the patient helps to evaluate the degree of anemia, the cardiovascular responses to fever and anemia, abdominal distension, localized tenderness, and the presence or absence of perineal disease as well as the eye, skin, and joint complications.

The simple laboratory studies of hemoglobin and hematocrit, white blood count with platelets, serum iron, sedimentation rate, or C-reactive protein supply me with the essential measurements needed to determine degree of clinical activity. We usually find that the endoscopic visualization of the colon is very well correlated with the clinical evidence of disease especially in ulcerative colitis, less so in Crohn's disease, bearing in mind that the use of rectal medications in the form of steroids, 5-ASA (or even cyclosporine in some recent studies) may substantially reduce the degree of local rectal and sigmoid inflammation. Using these determinations, it is not difficult to classify individual patients into the broad clinical categories of mild, moderate, or severe. Difficulties may and do arise in patients with milder symptoms whose endoscopic appearance and gross pathology are minimal. It is this not infrequent finding which makes me insist that my endoscopist do multiple biopsies even at the very first endoscopic assessment. Flexible sigmoidoscopy can be very helpful in establishing the proximal limits of disease in patients considered to have ulcerative proctitis or ulcerative proctosigmoiditis.

Clinical Assessment in Crohn's Disease

To turn to the more difficult problem of Crohn's disease, there exists no simple laboratory measurement in my opinion that facilitates this aspect. In general, the presence or absence of symptoms of fever, loss of appetite, abdominal pain, weight loss, and the addition of any of the group of extraintestinal manifestations of the "colitic group" leads us to infer the overall degree of impairment and act as a simple guide to label the patient as having mild, moderate, or severe disease activity and impaired function. But no single specific laboratory marker of inflammatory activity exists, and perhaps, in the end, our rapid synthesis of these general features in our patient is all we need at times. They are often all that we have to judge with.

It is in the absence of gross fever and other clinical features just mentioned that the physician most often wants to measure disease activity in Crohn's disease. The oldest means are by no means the best documented and certainly, in the United States, the most frequent used measurement is the *erythrocyte sedimentation rate*. The sedimentation rate is useful in colonic Crohn's disease in a very general way, but others, as well as ourselves, have noted that the sedimentation rate often does not help with small bowel Crohn's disease, a point that has been made by members of our group for some time now. Why patients sick with clear-cut, extensive small bowel disease should have sedimentation rates within the normal range is baffling and intriguing but real. Alongside the sedimentation rate, C-reactive protein and orosomucoids have long been championed and nominated as successful, useful measures of inflammatory activity. In a rough, general way, these measurements do measure the clinical severity of the Crohn's disease, but are not very closely connected to the location of the disease. Indeed, the *thrombocyte count* has also been stressed as a good independent marker of inflammatory disease activity. I have found the C-reactive protein sometimes useful in this context and only recently have been able to find commercial laboratories who can determine orosomucoid levels, but am not using it. However, I see little to choose from among these acute phase reactants.

I do note the very high platelet counts when they are available in my patients' hospital records. I know from studies done with my hematologic colleague, Dr. Louis Aledort, that these are normally functioning platelets and I am interested in these very high levels as possible indicators of ensuing venous thrombosis, especially the ileofemoral thrombosis, which has such an ominous prognosis in my experience in the young patient with IBD. Serum albumin reflects a complex of factors. It is also an indicator of intestinal mucosal leak of other serum proteins and thus serves as an indirect measure of inflammatory activity. Stool alpha-1-anti-trypsin, which is not digested by the proteolytic enzymes in the feces, we believe is a good measure of protein leaks in the colon and the small bowel mucosa and thus serves as an indirect indicator of inflammatory activity. Our studies on the concentration of this serum protein in random stool specimens in Crohn's disease has shown a fairly good correlation with the clinical assessment and other laboratory measures of inflammatory activity. It has varied with clinical improvement, and the concentration has fallen in keeping with clinical improvement and can be correlated with the use of therapeutic measures.

The assessment of the significance of abdominal pain in Crohn's disease presents the clinician with the problem of distinguishing the complex of inflammatory activity, local muscular spasticity, and mechanical stricture. Here we must use colonic endoscopy and biopsies, radiographic evidence of small and large bowel strictures along with laboratory indices of inflammatory activity. Often, only continued observation and the passage of time resolve the tangle of factors needed for a more accurate assessment of the degree of impairment of the patient.

Assessment of Nutritional and Absorptive Status

I discuss these two areas together because it is important and often difficult to separate malnutrition from malabsorption. In assessing the nutritional state of the individual patient, I am aware of the anthropomorphic measurements that the nutritionists use, such as the measurement of skin fold at critical areas of the body. But as a gastroenterologist, my approach is much simpler. Body weight judged against height without recourse to standard tables, noting abdominal wall wrinkling and pedal edema, are obvious evidence derived from physical examination. A routine laboratory study of hemoglobin and hematocrit, especially the red cell indices, measurement of serum iron, Vitamin B12, and folate levels are important in the assessment of nutrition. Levels of serum albumin and B12, iron, and folate give us some insight into the digestive and assimilative processes of the patient's intestinal tract. If we really need to more precisely measure malabsorption in a given patient, we can measure fat absorption by collecting 72-hour stool, measure carbohydrate absorption by doing a urinary or blood xylose test, and measure absorption of B12 by the Schilling test with or without intrinsic factor or antibiotics. We will need to make sure, especially in the case of fat absorption, that the patient can or will eat a high (100-gram) fat diet. So many of the 24-hour stool collections we see in our hospital charts are worthless because they do not have a measure of the patient's intake, nor can we rely on his or her information regarding this.

Often we want to know not only whether chronic bacterial overgrowth is present, since it may occur in any patient with either an area of stricturing or narrowing of the bowel or with an ileocolonic anastomosis, but we also need to know how important it is. Here the Schilling test of B12 absorption with or without intestinal antibiotics, or the hydrogen breath test following a lactose load can help firm up this diagnostic area from a quantitative point of view.

In assessing the state of disease in ulcerative colitis, rigid sigmoidoscopy, flexible sigmoidoscopy, and colonoscopy can give us the answer. Of course in involvement of the colon in Crohn's disease this too must be the case, but we will need to do a small bowel x-ray as well to determine the extent of disease. Yet our French colleagues have demonstrated that the endoscopic appearance in Crohn's colitis correlates poorly with the clinical manifestations. I believe that microscopic findings are better correlated with clinical activity. But, in addition to the geographical extent, the amount of normally functioning mucosa is important in assessing the factors involved in weight loss, diarrhea, osteomalacia with bone pain, and the formation of kidney uric and oxylate stones. Here the degree of ileal involvement is prominent. My rough measurement is the degree of visible involvement on radiographic studies, in other words, the measure of gross disease. Ileal biopsies through the colonoscope or jejunal biopsies via the oral route do not usually answer the question, but they may on rare occasion be helpful. In the functional studies of ileal function, the B12 serum level and the 72-hour fecal fat test are the more helpful modalities in this connection. The measurement of the fecal excretion of bile salts still remains a laboratory and research tool.

Adding Up the Assessment in Crohn's Disease

In determining what our therapeutic aim and goal should be before treating any one patient or else to assess whether our treatment is doing the patient any good or whether we are improving his or her lot, it would be convenient to total up an overview of these various modes of assessment—inflammatory, structural, and functional—in a simple numerical form. This overview is especially true whether we contemplate treating one patient or wish to compare treatment among groups of patients. I have already commented on the Crohn's Disease Activity Index (CDAI) of the U.S. national trials, but it unfortunately does not help to serve our daily practical needs. We end up with a global impression that an individual patient has a certain rough degree of disease, ranging from mild to moderate to severe inflammation in a given area of bowel with a rough measure of adequate functional mucosa or calorie absorption with or without some specific defects of absorption of defined clinical importance.

Immunological Assessment in IBD

So much of current research into the etiology and pathogenesis of inflammatory bowel disease focuses on the role of the patient's general and more recently the blood and intestinal tissue immunological status that we can't help wondering whether any such studies will help us in our overall evaluation of the patient. Unfortunately no routine immunological studies have much to offer us in assessing the patient's current impairment, particularly in predicting the response to medical therapy or the outcome of surgical intervention, although it has been suggested that the tissue content of immunoglobulins in resected specimens of Crohn's disease may have predictive value.

Skin reactivity is a poor measure of the individual's immune competence. Study of the subsets of lymphocytes is now the fashion, especially in patients with symptoms suggestive of inflammatory bowel disease who may be HIV positive. In studies at our own institution, the ratio of helper to suppressor T-cell activation in vitro appears to provide evidence of defective immunoregulation in Crohn's disease and in ulcerative colitis, but remains a research tool. Alpha-1-anti-trypsin simply measures leak from the plasma into the tissues in inflamed disease. It recently has been suggested that measurement of *tissue necrosis factor,* which is one of the many mediators of inflammation present in the stool, may also be correlated with clinical activity. It does seem closer to the biochemical chain that is involved in tissue inflammation than simply the permeability to alpha-1-anti-trypsin from the plasma, but this clinical use is still to be substantiated.

Circulating immune complex measurements unfortunately are not routinely available, but might be useful in trying to decide whether some of the general systemic reactions of the patient and especially the joint and other extraintestinal manifestations may represent a measure of the underlying inflammatory bowel activity rather than the symptoms of steroid withdrawal in our patients who have considerable joint pain.

Summing Up the Summary

So after all these areas have been considered, we can say that the patient before us has inflammatory bowel disease (ulcerative colitis, Crohn's disease, or indeterminate variety) in a mild, moderate, or severe form in anatomically defined areas of the gut documented by radiographic or endoscopic techniques with roughly defined defects of absorption (calories; vitamins, especially folic acid and Vitamin B12; minerals, especially iron) with crude measurements of the nutritional impairment based mainly on weight and the degree of anemia and hypoalbuminemia complemented by overall global estimate of degree of impairment in the particular individual's daily life functioning. It is clear to me that no single numerical index, no one number can replace this relatively crude assessment of the degree of the sufferer's illness. Yet crude as it is, it is the best we have. We need this assessment in our outlook to predict our patient's outcome and to plan our therapies to achieve our stated goals.

This may be an appropriate place to consider the "stated goals" just referred to. Since none of our therapies except colectomy for ulcerative colitis "cures" these disorders, we obviously aim at suppressing the patient's symptoms and in improving the quality of their lives. It is conventional to consider our goal as the induction of a remission, but it may be very important in some instances, especially in Crohn's disease sufferers, to define our therapeutic goals more specifically. This Present and Korelitz did in their original 6-MP study and this in turn has been used by some others. The goal in any one individual may be to heal a fistula or be able to wean completely from steroid medication or return to an active career for another.

13

Prognosis

Predicting Outcomes and Reading the Future

Of all of the parts of the art of medicine, Hippocrates said that prognosis, predicting the outcome, was the hardest. This statement remains especially true of IBD.

When we make the first diagnosis of any of the variants of ulcerative colitis, from proctitis to left-sided disease to universal ulcerative colitis, or attach the definitive label of Crohn's disease to a young or even an older person, what can we say about the outcome? Torn between our need to reassure our patients and their loved ones and our recognition that we have few firm facts to go on, our glimpses into the future for any one individual are indeed cloudy.

Predicting the Future in Severe Ulcerative Colitis

The outlook for survival of a patient with severe ulcerative colitis, however defined, has improved remarkably in my own lifetime in IBD. In the 30 years since the introduction of steroid therapy for ulcerative colitis, the mortality of this variant has declined remarkably. This has been dramatic and is well established. When my colleagues and I reviewed the current medical therapy of severe ulcerative colitis from five comparable clinical trials, including one of our own—well-defined prospective studies with severity defined in all by the criteria of Truelove and Witts—there was no reported fatality among 295 patients treated between 1974 and 1990! This clearly has been the result of our improved anesthesia, intensive preoperative care, and surgical skills.

Achieving clinical remission while still preserving the colon, however, has not been so successful despite the claims of several retrospective studies. Of all 295 subjects just alluded to, remission was achieved in 53 percent (95 percent confidence interval = 49 to 79 percent), the lowest remission in any subgroup was only 25 percent. In the 205 treatment courses in which colectomy was the predetermined endpoint of medical failure, 35 percent of the subjects lost their colons.

The question that we all would like to answer is: Can we predict from any feature of the clinical picture those individuals who will be expected to respond to current therapy and keep their colons?

It has been suggested by a few prior investigators that patients presenting on admission to a hospital with severe ulcerative colitis with a fever of above 101°, nine or more stools in 24 hours, serum albumin below three grams, and tachycardia would most likely leave the hospital without their colons.

When we looked at the measurements we made in the course of trials of medical therapy in our 66 patients with severe ulcerative colitis mentioned already, we included approximately 30 determinations of both laboratory and clinical observations made on admission and daily thereafter for 10 days. No single factor or combination of factors had any reliable predictive value. In most studies, transfer from another institution to the reporting one and toxic dilation of the colon carried with it a slightly worse outlook and my personal experience is consistent with this generalization.

What can we tell our sick patients if they go into remission in the course of an acute severe attack of the chances of remaining well even if only after the first six to 12 months after successful therapy? Not too much which is reassuring. They will have a tendency to remain well for six months in 54 percent of the cases (95 percent confidence interval = 45 to 63 percent) and only 35 percent of those will remain well at the end of the first year (95 percent confidence interval = 32 to 39 percent).

What is particularly frustrating in this situation is that after these last 30 years of observation by experienced clinicians, we still lack information of what effect the prior ulcerative colitis history plays in the rate of recovery of severe disease. There are available no data on the effects of the length of prior illness, number of episodes in either mild, moderate, or severe attacks, and the extent of the colonic involvement that bear on the likelihood of response during the presenting episode. My belief is that the therapeutic response to the first attack is more likely to be more successful than the response in the subsequent episodes, but there is only little objective evidence to support this belief.

Predicting the Future in
Mild to Moderately Severe Ulcerative Colitis

Turning to ulcerative proctitis first, the most important statement I can make to the patient is that it is most unlikely that the localized rectal disease will extend throughout the entire colon, but there is increasing uncertainty in my mind, if not in my voice, that one is confident that no more than 5 percent of the patients will have the process extend. My increasingly recent experience has been enlarging this number. Ten to 15 percent seems more realistic. This information is consistent with the studies of Powell-Tuck and his colleagues who used an actuarial method of analysis. Proctitis extends within five years to the sigmoid in about 10 percent of patients and to 20 percent in 10 years and to 30 percent in 20 years in their experience. The risk of extirpative surgery was only 2 percent in five years and 5 percent in 10 years. The risk of cancer in proctitis is probably no different than in the general population. These reassuring features of proctitis have in general been derived from radiographic and rigid sigmoidoscopic reports. They need to be amplified by the increasing information to be derived from total colonoscopic biopsies in patients with localized proctitis. This already appears to indicate that microscopic involvement of the left colon may exist and extend beyond the gross appearance of the rectal involvement. Equally important in discussing outcomes with patients we will look with interest to see whether our newer local therapies really improve the long-term outlook in proctitis.

Very important also in the first discussion with the patient is the desirability of firmly stating that the incidence of rectal cancer is not increased in ulcerative proctitis. Further, life expectancy is generally not altered. However, to the real question of the risk of recurrence and its severity, we can say little except that this variant does recur and relapses of unknown origin are the rule, encouraging the patient to try to discern if any pattern exists in the circumstances in which the recurrences do recur in his or her lifestyle. Cautioning against the too liberal use of wide spectrum antibiotics, as well as the nonsteroidal anti-inflammatory agents, is important.

Prognosis in General in Ulcerative Colitis

No matter how great our reluctance to discuss the long-term outlook of individual patients with ulcerative colitis we may be in our first contact with them or in the course of the first episode, sooner or later we shall have to point out, however gently we do, that the susceptibility to recurrence is a lifelong affair eliminated only by eliminating the colon.

The Outlook of an Individual Attack of Left-sided or Universal Ulcerative Colitis of Mild to Moderate Severity

It is difficult to assure the anxious patient with an episode of mild to moderate ulcerative colitis of universal or left-sided extent as to his or her chances of responding to our current therapy, but we all feel confident, at least during the first attack, that the patient will sooner or later respond to our current therapies.

To obtain a little further data in this regard, my colleagues and I have looked at some 11 published trials of mild to moderate severity ulcerative colitis involving 232 patients receiving a single drug therapy under well-defined, controlled conditions. If one insists on complete clinical recovery as the endpoint, 60 percent recovered. The range was from 24 to 90 percent. If partial improvement was added to complete clinical recovery, this criterion improves the favorable outlook to about 70 percent, ranging from 65 to 100 percent. As we all know, sigmoidoscopic remission lags behind clinical remission, averaging in these patients 46 percent (the range was 30 to 60 percent). However, we all also know that in the real world of clinical gastroenterologic practice, if the patient with mild or moderate ulcerative colitis fails in one form of treatment, a second or a third drug are added successively or synchronously.

What can we tell the patients who have remitted or will remit about their chances of remaining well for the next six to 12 months even if they continue on the successful drugs which induced their improvement? In the review of the just cited material, 176 patients with localized or universal ulcerative colitis of mild to moderate severity were maintained on the single drug that had induced the remission for from six to 12 months. Seventy percent could be reassured that they would continue their remission (range 38 to 80 percent) for at least six to 12 months.

If the prognosis of a single episode of this severity of ulcerative colitis is good, what can be predicted regarding the recurrent course? Here we are dependent on the carefully collected material of our British and Scandinavian colleagues with their complete follow-up studies of a defined geographical area. In Copenhagen county, for example, a relapse-free course was found in only 3 percent of patients after 10 years and in less than 1 percent after 18 years. In any one year after the first three, however, about half the patients appeared to be in remission and an intermittent course was found in 90 percent of the patients treated medically. Similar results regarding intermittency have been reported from England and Scotland. It would seem to be true that patients will be symptom-free for a little more than half of the time. This prediction seems to be the case in all patients we see, whether they have rectal or left-sided or universal involvement. We can further reassure our female patients that, in the modern era of medical and/or surgical treatment, their survival with ulcerative colitis is really no different from the remaining population, and this principle also holds true for men except for a slight excess mortality in the first two years of the illness.

All patients with ulcerative colitis, regardless of their degrees of sophistication, fear coming to surgery with its threats of ileostomy, despite our newer, more cosmetic, rectal-saving operations. What proportion of patients with ulcerative colitis eventually come to operation? Data from tertiary referral hospitals, such as my own, cannot help answer this question. Better demographic data of such areas as Scandinavia and Great Britain, with their good public health records, suggest perhaps 20 to 25 percent in 10 years and 30 percent after 15 to 18 years, which fits in with my general impression.

Prognosis in Crohn's Disease

Predicting the outcome of Crohn's disease is even more difficult than in ulcerative colitis, complex as that disease's course may be. In all categories treated medically with our current armamentaria (steroids, sulfasalazine, antibiotics, including metronidazole, newer variants of sulfasalazine, immunosuppressive drugs of the azathioprine or 6-mercaptopurine group) we can safely say that Crohn's disease is lifelong with unpredictable exacerbation or remissions and that the concept of "cure" has no real meaning at present however much we improve the quality of our patients' lives. Ye the risk of cancer, discussed more fully in Chapter 26, is real, but not as great as in ulcerative colitis, and I see no need to emphasize this aspect early in the management of newly diagnosed cases, especially in youngsters with localized ileal disease. Late in the course, especially in those with extensive colonic disease, it will have to be faced.

Yet we all have patients who have gone into remission with whatever medical therapy or operation they received and have remained in remission for periods as long as 10 to 30 years. At the same time, in discussing recurrence after treatment, it should be borne in mind that almost all patients will have some histologic evidence of recurrent disease in the first year after the operation. Many will have radiographic evidence of recurrence, but the important question is whether or not they have significant clinical recurrences, including extraintestinal manifestations and especially perianal disease.

In an effort to obtain some data regarding the likelihood that our patients with Crohn's disease were better off after a reasonable course of our current medications, my colleagues and I looked at the reported data and some well-controlled trials of a single drug against placebo in mild and moderate Crohn's disease patients however defined using the Crohn's Disease Activity Index (CDAI), the Harvey–Bradshaw Index, or the Van Hees Index and who were well enough to be treated in an outpatient setting. In six trials of 332 individuals treated with a single drug for six to 26 weeks, 49 percent (range 31 to 51 percent) went into complete clinical remission defined by these indices. If one considers six others in which the endpoints were complete and/or partial improvement, the chance of success was somewhat better: 53 percent (range 40 to 70 percent).

In an attempt to answer our patients' question if they would remain better after therapy, we looked at eight well-controlled trials of 470 patients maintained on the medications, which induced their remission 40 percent (range 20 to 71 percent), who remained improved for periods of 12 to 36 months, except the one study on azathioprine maintenance in which the remission rate remained close to 95 percent.

This kind of analysis gives only fragments of needed information. In the real world of gastroenterologic practice, patients are given successive and often synchronous medications if they fail one group.

At the end of the section in Chapter 5 on the varieties of clinical presentation in Crohn's disease, I alluded to the emerging concepts of at least two forms of Crohn's disease—a "perforating" and a "nonperforating" one. The rough distinction carries with it some rough prognostic implications. Not only does the perforating form come to operation earlier than the nonperforating form, but recurrences take place earlier as well. Since the future is not identical with the past, but tends to resemble it, it is not surprising that the form in clinical presentation of recurrences is generally similar to the original presentation in each variety. However, more long-term prospective follow-up studies are needed, but a retrospective look at a large group of patients with CD at our own institution is interesting. Some 770 patients who underwent resection during the years 1963 through 1990, the indication for a second resection closely followed the indications in the first operation. Of those who were operated on for "perforating" indications, more than 70 percent had the same indication for reoperation. Similar findings held true for those operated on for a "nonperforating" indication.

In summing up, if patients come to operation with a perforating indication (fistula, abscess, free perforation) rather than a nonperforating indication (obstruction, intractibility, and hemorrhage), they are likely to come to operation again for the same reason as was the case in the first operation. However, patients are not likely to come to reoperation for perforation much earlier than if their indication for surgery was nonperforating, judging from more recent studies.

Prognosis of Toxic Dilation in IBD

In discussing the outlook for patients with severe IBD, the dramatic presentation when toxic dilation of the bowel is present, because of its grave threat to life, deserves separate consideration, although the incidence is rather low. Of a little more than 1200 patients with IBD distributed equally between ulcerative colitis and Crohn's disease, 75 patients (about 6 percent) had this presentation at our hospital between 1960 and 1979. Although some of our close clinical colleagues have had much better luck with patients, this complication is extremely serious: 12 of our patients (16 percent) died. So what can we predict about our next new series of patients with toxic dilation? What are the odds of surviving this disorder, coming to emergency surgery, and the long-term outlook of these patients if they survive without operation?

Although dilation occurs more frequently in ulcerative colitis than in Crohn's disease (10 percent versus 2.3 percent), it is reasonable to consider this group as a whole as they behave similarly. We may assure ourselves and our patients' families that at least 90 percent of patients have survived this complication, whether treated medically or surgically. It was better to be under 40 years of age and much worse to have had a perforation, and it didn't make much difference whether the present episode was the first attack or one in a more chronic course or whether the disease was confined to the left colon or involved the entire colon.

Almost 80 percent of these patients required urgent colectomy. About 80 percent of them survived. Of the 20 patients who were treated mainly and solely medically, over 85 percent survived.

What about those who survive the acute presenting episode, what is their long-term outlook? In our group of 14 patients treated without operation, eight (or 57 percent) came to subsequent colectomy, two having had recurrences of toxicity or dilatation. This kind of long-term outlook is in keeping with the reports of large groups of patients treated in other referral centers, such as the Mayo Clinic. Even those clinicians whose mortality and surgical intervention rates have been lower than our collective experience also found that half of their patients come to colectomy in the long run.

VI

MEDICAL MANAGEMENT

14

The Natural History of Inflammatory Bowel Disease

The Lessons of the Placebo

We need to know more about the natural history of both inflammatory bowel diseases in adults in order to answer the most primitive question: Must every IBD patient be treated? The question, as far as I can recall, has been a rhetorical one ever since these disorders were first defined. No patient suspected of either ulcerative colitis or Crohn's disease remains untreated at present.

Yet the emergence of the category of acute self-limited colitis certainly raises the question whether ulcerative colitis and Crohn's disease can have a self-limiting course. Since we all believe that Crohn's disease and ulcerative colitis are lifelong disorders, it is only the occasional appearance of a patient with what looks like ulcerative colitis who recovers and remains well thereafter that has raised the possibility of self-limited acute colitis as a separate entity.

This remains as a diagnosis after the fact, although some sophisticated pathologists believe that characteristic microscopic features of colonic or rectal biopsy tissue can be used to differentiate this group from the better known inflammatory bowel diseases. I see no reason to gainsay this point of view or deny the existence of the entity of acute self-limited colitis. Since we cannot easily predict the outcome of a single first episode, all of our suspects will continue to be treated with the usual drugs and such is my procedure as well. I suspect that the self-limited intestinal inflammations probably represent an acute, specific viral or bacterial infection. Indeed, if the colonic or ileal lesion seen endoscopically or radiographically clears completely and remains healed indefinitely, I am reluctant to label the patient as having had either ulcerative colitis or Crohn's disease. I go so far as to think that any ileal lesion that clears completely radiographically probably was not Crohn's disease, but more likely a *Yersinia* infection or related to acute appendicitis. One can't help wondering what possible role C. *difficile* toxin might be playing in these sporadic instances of self-limited acute colitis.

This is not a chapter about the placebo response, nor is it a plea to treat IBD with placebos. However, if we wish to predict the response of any group of patients with these disorders to our current therapies, we need to know a good deal about the natural history of these disorders themselves. Unfortunately, we have little secure information about untreated ulcerative colitis and less about Crohn's disease.

As regards ulcerative colitis, including the milder variants, especially proctitis, little is known about their untreated natural history. Few studies of a controlled rigorous nature were done before the steroid era. The early studies of the 1950s on cortisone in ulcerative colitis led British workers to conclude that this drug was useful, although the number of controlled trials were few. Indeed, the current climate of opinion is so hardened that it would be considered unethical for a physician not to treat the patient sick with ulcerative colitis with the available drugs and especially steroids, so we are unlikely to have further controlled trials.

As for Crohn's disease, the situation is even worse. We need to know a good deal about its natural untreated history if we are to estimate the value of any of our medicines in this disease that has such a variable and protracted course.

The only possibility of obtaining such needed information, I have long argued, is to use with discretion the placebo group of controlled trials such as do exist. I am well aware of the biases that exist in such studies and the selection of patients sick enough to be treated yet well enough to be followed for weeks, months, and even years. In addition, patients in such clinical studies are really not untreated and may not truly portray the "natural" state. The active participation in this kind of study, being thought capable of understanding and complying, then being seen frequently by doctors and other members of the health care team, keeping careful records, and taking medications believed to be potent make an undefinable difference, yet these placebo patients are our only source of needed information. So we must use this material. A particular study may not provide us with all the clinical details we would like to have, but, by looking at the entire group of studies, important questions can be asked. Can a patient sick enough to need treatment be better off without any specific therapy and how often can this occur? If the patient is in remission, what are the chances of his or her continuing to do well? Are there any factors capable of predicting who will achieve or maintain remission?

The Natural History of Ulcerative Colitis

In an early analytic study of placebo-controlled trials in ulcerative colitis, Samuel Meyers and I accepted 12 studies of 185 patients with active disease; only three had severe disease. The majority were mild to moderately ill. Clinical and sigmoidoscopic improvement were the end point in the trials that lasted from 14 to 44 days. In these studies of those on placebo alone, 30 percent improved clinically (range 16 to 52 percent) and even 38 percent improved sigmoidoscopically (range 26 to 59 percent). Relapses, however, occurred commonly within two months of the spontaneous improvements. These studies were done in the years 1958 to 1987. We accepted six similar studies of maintenance therapy on placebo in 164 patients: 134 followed for six months and 30 for 12 months. Of the six-month trials, 41 percent remained in remission (range 20 to 51 percent) and 25 percent remained in remission for 12 months without any specific medication. These studies were done during the years 1965 to 1972. More recently, my more rigorous young colleagues, Asher Kornbluth and Peter Salomon, persuaded me that we should look at 11 studies of 236 patients with mild to moderately severe disease treated for periods of 14 to 42 days where complete remission was separated from partial clinical improvement and sigmoidoscopic remission rather than improvement in one grade was the end point.

In this group of subjects data on complete clinical remission was available in seven studies of 126 patients. Ten percent achieved complete remission (range 0 to 17 percent) and, of the patients who achieved complete remission or partial remission we looked at, the overall proportion of the total analyzed achieving either level of improvement was 30 percent (range 18 to 43 percent). Complete sigmoidoscopic remission was achieved in four studies of 50 patients in only 8 percent. These studies were done between 1958 and 1989. We also accepted five studies of placebo maintenance trials of 167 mild to moderate ill patients. In the four studies of 134 patients followed for six months, 51 percent remained in remission (range 40 to 71 percent), and of the 33 patients followed for 12 months, 24 percent remained in remission. Thus, of our mild to moderate sick patients with ulcerative colitis, some 30 percent may improve spontaneously and an equal number may have an equal improvement in their sigmoidoscopic appearance within two to six weeks, although only 10 percent (ranging up to 17 percent) will achieve a complete clinical remission and 8 percent complete sigmoidoscopic remission in the same two to six weeks.

If patients who had achieved a remission in their mild to moderately severe ulcerative colitis by whatever means can remain in remission upon no specific medication for 40 to 50 percent for six months, and for 25 percent for 12 months, predicting which specific patients will remain in remission still remains uncertain.

The Natural History of Crohn's Disease

To achieve some perspective in the tendency of Crohn's disease patients to improve simply with supportive care, but without any anti-inflammatory drugs, Samuel Meyers and I looked at three controlled studies of placebo-treated patients with mild to moderate Crohn's disease who could be managed as outpatients. Of some 121 subjects treated for 16 to 26 weeks, 25 percent remitted (range 8 to 42 percent). Looking at the question of maintenance of remission of some 199 patients receiving only supportive care as outpatients, it was clear that 48 percent continued their improvement at 12 months (range 21 to 66 percent) and of the 138 patients in remission 38 percent remained well for two years (range 35 to 40 percent).

When Kornbluth, Salomon, and I looked at this kind of study more recently, in 11 trials of placebo therapy for six to 26 weeks in 442 patients with Crohn's disease, we observed that 27 percent (range 8 to 36 percent) achieved complete clinical remission, whatever index was used to define this degree of improvement, and it increased to 31 percent (range 8 to 50 percent) when those with partial improvement were included as well.

When we more recently reviewed placebo-controlled maintenance studies in Crohn's disease without inflammatory drugs in some 583 patients followed from one to three years, 41 percent remained in remission (range 28 to 64 percent). However, when we looked at specific time intervals, 64 percent remained in remission for one year, which declined to 46 percent at 1.5 years, to 38 percent in two years, and to 28 percent at three years.

So looking at this catalogue, we concluded that about 25 percent of Crohn's disease patients with a mild to moderate variety can improve or remit spontaneously when observed over periods of six to 26 weeks, and once in remission up to 40 percent can remain well for two years, and even 28 percent for three years.

15

Medical Management Programs

General Drug Considerations

In discussing treatment of inflammatory bowel disease, the word "cure" rarely enters our lexicon since the causes of ulcerative colitis and Crohn's disease are unknown. Ulcerative colitis almost always can be "cured" by resection of the colon, but at the cost of either a life with ileostomy or one with the problems associated with "pouchitis," of the abdominal pouch of Kock or the pelvic pouch. Crohn's disease we know is lifelong and both ulcerative colitis and Crohn's disease may go into remission and remain in remission for significant periods of time as both our patients and the placebo-controlled trials tell us. So what can we expect and our patients hope for from our present day management?

Thus, in assessing the value of our treatment methods, we regularly distinguish between *acute* treatment and *maintenance* treatment and know and believe that our current drugs do not attack the etiologic basis of the illness. The consensus of workers in the field is that we are treating the effector arm of the inflammatory bio-chemical–immunological chain, and I have always held that maintenance therapy is essentially the same as acute phase therapy. In both phases we are attempting to suppress the same effector chain. Maintenance therapy is an attempt to suppress and minimize or attenuate mini-episodes of disease, mini-flare-ups, if you will. Yet it is true as Derick Jewel has recently pointed out: "There is no evidence of an a priori reason that treatment of acute disease and prevention of relapse occur by the same mechanisms." But if we are not getting at the fundamental causes or triggering mechanisms in inflammatory conditions, it has always seemed reasonable to me to expect that any medication that could induce a remission should be able to maintain that remission and, in general, this expectation has been held true for most of our drugs. It has always puzzled me and been a stumbling block in our management programs that sulfasalazine, which is valuable in treating the acute episodes of Crohn's disease, has not been demonstrated conclusively to protect patients during maintenance trials. So I have felt that the emerging evidence that the newer 5-ASA compounds can favorably improve the maintenance phase in Crohn's disease and support long-term use of these drugs is unifying our thinking. But, in a decision to continue any acute phase drug into the maintenance phase, the calculation must take into very serious consideration in its long-term use the adverse consequences of any of our medications. I think this reasoning regarding the similar mechanisms of acute and maintenance therapy may hold as well for some of our newer drugs of the immunomodulating variety.

The Available Drugs

As the background for my discussion of current drug therapy and IBD, I have sketched the available evidence for spontaneous improvement and its maintenance from the placebo studies of controlled single drug acute and maintenance studies. But, when we turn to the day-to-day management of patients, we are faced with an embarrassment of riches (see *table*). We have a whole host of drugs to choose from and current therapeutic fashions lead us to add one drug on to another when the patient's course is not improving.

Available Drugs for IBD

Steroids

Oral	Topical	Parenteral
Prednisone	Hydrocortisone hemisuccinate	Hydrocortisone
Prednisolone	Hydrocortisone acetate foam	6-methylprednisolone
6-methylprednisolone	Prednisolone-21-phosphate	ACTH
	Declomethasone dipropionate	
	Betamethasone	
	Toxicortical pivalate	
	Prednisolone metasulfabenzoate	
	Budesonide	
	Fluticasone propionate	

5 Aminosalicylates

Oral	Topical
Sulfasalazine (Azulfidine®)	5-ASA suppositories
5-ASA	5-ASA enemas (Rowasa®)
Mesalamine (Asacol®, Pentasa®, Claversal®, Salofalk®)	
Olsalazine (Dipentum®)	
4-ASA (PASA)	4-ASA enema

Antibiotics

Metronidazole (Flagyl®)
Ampicillin, tetracycline, ciprofloxacin (Cipro®)

Immunosuppressants

Azathioprine (Imuran®)
6-mercaptopurine (Purinethol®)
Methotrexate
Cyclosporine A (oral, parenteral, rectal)

Candidate Drugs

Hydroxychloroquine (Plaquenil®)
Omega 3 fatty acids
FK506
5 lipooxygenase inhibitors
IL-1 receptor antagonists
CD4 monoclonal antibodies
ICAM inhibitors
Interferon-alpha
Lidocaine

I propose next to comment briefly on the many medications we now have available and then to turn to the diseases themselves and indicate the order in which I believe it is reasonable to use them, discussing as always ulcerative colitis separately from Crohn's disease.

The Group of 5 Aminosalicylates

Sulfasalazine, the combination of 5 aminosalicylate (mesalamine) and its carrier, sulfapyridine, is certainly the oldest drug and the most widely used medication in the treatment of IBD. Synthesized at the suggestion of the late Nana Swartz of the Karolinska Institute for the treatment of rheumatoid arthritis in 1942, she astutely noted its therapeutic effect in patients with ulcerative colitis in which disease it is now the standard form of therapy. It is interesting to gastroenterologists that sulfasalazine has been prescribed in recent years by rheumatologists for the treatment of arthritis, its original goal. (Paralleling this exchange has been the adoption of the idea that the rheumatologists' success with hydroxychloroquine might be useful in ulcerative colitis.)

Sulfasalazine finds its clearest usefulness in the treatment and maintenance therapy of mild and moderate ulcerative colitis. It is less effective in Crohn's disease in inducing remission and its role in maintenance therapy is controversial and evolving, however, I favor its use in this disorder.

When one considers the vast amount of sulfasalazine prescribed and taken by patients in the past 50 years, it must be considered one of the safest drugs used in IBD. Mild intolerance, in the form of slight nausea, slight anorexia, and headaches, in my experience, is common when the drug is given in large doses initially and can be managed by starting at lower doses, and responds to antihistamines.

Severe intolerance conventionally ascribed to the sulfa moiety, especially in patients who are slow acetylators, includes vomiting, rash, and fever. This group of side effects includes as well neutropenia, rarely agranulocytosis, and acute hemolytic anemia (usually Coombs negative) presumed to be related to the sulfasalazine level and acetylator activity. Although it has been long known that sulfasalazine interferes with folate absorption and may lead to a megaloblastic anemia, many physicians fail to prescribe the obligatory daily 1 mg of folate. Yet the occurrence of this complication must be rare. I have seen it only once and is probably ascribed to the fact that, in my assumption, most patients with IBD are taking some form of multivitamins which contain varying amounts of folic acid.

In my experience, puzzling, obscure side effects include rare pancreatitis, pleuritis, and pericarditis. Even more rare is a lupus-like syndrome with even positive lupus preps. Pulmonary fibrosis is extremely uncommon, but a persistent, undiagnosed cough was present in several of my patients, diagnosed in one, when a supply of the drug during a trans-continental trip ran out and the cough disappeared as well.

Suppression of the male sperm count must be rare, and it took many years before it was detected or even suspected and clears when stopping the medication. My advice to the husbands on sulfasalazine is to switch to other forms after a trial of attempting to impregnate their spouses fails, but some prefer to stop or change to the newer 5-ASA preparations before the attempt to impregnate their partners.

A more trivial, but occasionally disturbing, side effect in a number of members of both sexes is hair loss. I believed for a long time that the catabolic effects of the disease and the use of steroid preparations were the main reasons, but some more recent experience has convinced me that sulfasalazine can also lead to hair loss. I have not seen it so far in those in the newer 5-ASA forms, but others have; the period of observation of these drugs has been rather short.

For the time being, I have seen no advantage for my patients doing well on sulfasalazine to switch to the mesalazine forms, except in the presence of significant intolerances. I have noted that some patients, aware that sulfasalazine contains a sulfa portion, believe erroneously that they must drink large amounts of water to prevent kidney damage. Continuing statistical evidence exists that sulfasalazine can be used safely in pregnancy and nursing women. I discuss the question of a woman's attempt to become pregnant while on the drug in Chapter 25 on pregnancy and IBD.

Sulfasalazine-Related Compounds

ORAL PREPARATIONS

The elegantly simple experiments of Sidney Truelove and his colleagues have clearly shown that the active principal of sulfasalazine is the 5-ASA (5 aminosalicylate) component. This, in turn, has led to a flock of 5-ASA preparations worldwide that now have U.S. FDA approval. These include Asacol®, Pentasa®, and Dipentum® which contain 5-ASA, with a variety of coatings designed to release the active principle in the distal bowel by virtue of local pH changes or time released and Dipentum® (olsalazine)—a dimer of mesalamine split by bacterial action.

The obvious advantage of mesalamine preparations is their use in patients sensitive to sulfasalazine, an expectation that so far has borne out very well in my experience. In the relatively short time these variants have been available, I have seen only one or two individuals who were intolerant to both sulfasalazine and mesalazine, but I have not seen any of the major side effects in sulfasalazine in those of 5-ASA. This statement must be qualified by the fact that, in our own double blind or randomized studies of ambulatory patients with mild to moderate ulcerative colitis, olsalazine, the dimer of 5-ASA, caused diarrhea in 10 percent of the patients. In the short course of current clinical use, I have seen one example of thrombocytopenic purpura following exhibition of one variant of 5-ASA. Others have also been reported, as well as an episode of pancreatitis.

I have seen no renal damage in a limited use of the maximal dose of eight tablets so far. The significance of some formed elements of the urinary sediment in doses of Asacol® at 4.8 grams/day reported by others needs further confirmation. Recently two patients have been reported as having a myocarditis—one fatal while on a preparation of mesalamine.

I have no experience with the 4-ASA, oral or topical form of variants (paraamino sulfacyclic acid), which has been reported as efficacious in mild to moderate ulcerative colitis in both acute and maintenance studies.

TOPICAL 5-ASA

The discovery of the therapeutic effectiveness of 5-ASA by mouth in ulcerative colitis led to its prompt use in the topical form of rectal installation. The often dramatic response in ulcerative colitis and ulcerative proctosigmoiditis has resulted in its widespread adaptation. In the United States, Rowall's preparation, Rowasa® enemas (4 grams of mesalamine), is easily available and accepted well by patients. It is rare for patients to have any unfavorable side effects. A few of my patients have complained of some rectal burning on irritation possibly related to the preservatives used to maintain shelf life. The rapid response in some patients has led them to prematurely discard the use of this mode of administration. Available at present are 5-ASA suppositories, in 0.5- and 1-gram doses for use in acute rectal IBD.

Steroids

Since the initial report of Dearing and Brown of the use of cortisone and corticotropin (ACTH) in ulcerative colitis, and reinforced by the earliest controlled trials of cortisone by Truelove and Witts in 1955, corticosteroids have been used widely and often as the first choice in this disorder. But much needs to be learned regarding their appropriate place and dosage in their role. All physicians are aware of the serious and at times catastrophic side effects of steroids. Here I only wish to state my firm belief that prolonged use of steroids by any route is not an acceptable form of treatment. I see no place for maintenance therapy with prednisone or its congeners. That does not mean that I have not seen many patients who have become dependent on small doses of steroid who will not give them up even with a history of recurrent flares on steroids or as they were being weaned to lower doses.

There is good evidence that there is a place for corticotropin (ACTH) in the treatment of ulcerative colitis, especially in those seen sick with severe disease and have not been on recent steroids within the last month. Although a few patients have developed adrenal hemorrhage during the seven- to 10-day course of IV ACTH (120 units, 24 hours), this must be rare since I have not seen it in my extensive experience. Perhaps a routine abdominal sonogram of the adrenal area should be done in all patients about to receive this form of therapy. All the suspected cases have recovered without difficulty or residual impairment of adrenal reserve.

NONABSORBABLE STEROIDS

Ideally, we need steroid preparations that are not absorbed so as to have no side effects and no long-term impairment of the pituitary adrenal axis. Of the drugs listed in the table, only fluticasone achieves this by being cleared by the liver in the first pass through that organ. Prednisolone metasulfabenzoate and prednisolone-21-phosphate work by poor intestinal absorption. None of the others are available for oral use, although a slow oral release form of budesonide is under trial.

TOPICAL STEROIDS

The use of cortisone derivatives in the form of rectal enemas has been long established as a useful procedure in dealing with diseases of the left colon, but the usual dosage of hydrocortisone enemas, 60 mg, leads to considerable absorption. Long-term maintenance has some of the undesirable side effects as in long-term oral use. There is also available topical steroid foam preparations the patient can use during their active day (Cortifoam®, Proctofoam®) which have a lesser amount of steroid.

TOPICAL NONABSORBABLE STEROID PREPARATIONS

Of all of the older established drugs with fewer systemic effects, declomethasone dipropionate, betamethasone-17-malonate, and prednisone metasulfabenzoate have been demonstrated to be useful in a small number of trials, but are not available in the United States or United Kingdom. Of the new ones, toxicortical pivalate probably will not be released soon, if ever. Budesonide, being studied intensively in Sweden and now in the United States, seems most promising and we look forward to its availability, but this seems some time off.

A recent interesting approach to fashioning an oral preparation of a steroid with little systemic side effects has been the coating of prednisolone preparations, prednisone betasulfabenzoate with the acrylic resin Eudraget S, similar to the fashion of preparing oral 5-ASA preparations; this compound is now receiving clinical trials and may furnish a steroid preparation with minimal side effects.

Antibiotics

In the absence of any clear-cut microbiological agents in IBD, the justification for antibiotics is truly empirical. Any demonstrated suppurative complication certainly calls for and deserves parenteral antibiotics directed against the usual intestinal inhabitants although we are all aware of the role of C. *difficile* toxin supervening in their clinical course. Like many other investigators, I consider toxic dilation of the colon in either ulcerative colitis or Crohn's disease as a septic complication—that is, extension of the disease process in the mucosa or submucosa through the entire wall of the colon—and advocate use of antibiotics with wide intestinal coverage. Recently, their use in severe ulcerative colitis, even in the absence of toxic dilation, has been reemphasized. This has been the custom for some time in some standard defined trials and therapy of severe fulminant colitis.

Metronidazole (Flagyl®) is of interest because, in addition to its antibiotic activity against a wide variety of intestinal organisms, especially anaerobic, it has anti-inflammatory activity. While its place in the therapy of ulcerative colitis is debated and not convincingly demonstrated, it has won a considerable following in the treatment of Crohn's disease since its introduction in 1975 by Ursing and Kamme, especially in the treatment of perineal disease, abscesses, and fistulae with ulceration.

Although a metallic taste, mild nausea, and some cramps are well tolerated by patients as these side effects settle down, the neurogenic ones of tingling, numbness, and neurogenic pain are more troublesome. While these too will disappear or subside with either a reduction in dosage or cessation of the drug, they may persist for a long time despite the discontinuation of the medication and may in some rare cases prove irreversible. I see a place for Flagyl® in many varieties of Crohn's disease, but none in instances of ulcerative colitis.

Other Antibiotics

Sulfathaladine, a sulfonimide with little or poor intestinal absorption, not available at present (although patients have stockpiled supplies for themselves), was widely used by the pioneer generation of gastroenterologists in IBD (Crohn, Bockus, Bargen). At present, antibiotics such as ampicillin, tetracycline, and ciprofloxacin are widely used in Crohn's disease without convincing trials. It is sad that millions of dollars spent on countless clinical trials in IBD have produced virtually none on antibiotics.

Immunosuppressant Drugs

AZATHIOPRINE AND 6-MERCAPTOPURINE

The major change in my thinking of management of IBD during the past decade has been the acceptance of the value of these immunosuppressant agents. Azathioprine has a different side chain from 6-mercaptopurine, which is metabolized in vitro to 6-mercaptopurine, and I consider them interchangeable. Introduced for the therapy of ulcerative colitis long before its advocacy for Crohn's disease by Bryan Brooke in 1968, its place in ulcerative colitis fell off dramatically in part because of its carcinogenic possibilities in this disorder with its own considerable risk of cancer plus the recognition that surgical extirpation in the colon "cures" ulcerative colitis.

However, there is still a place for these two compounds in ulcerative colitis—in patients not sick enough to be considered as candidates for surgery, yet unable to be weaned from steroids. My anxiety about these carcinogenic potentialities having been overcome, I see a definite place for azathioprine and/or 6-mercaptopurine in Crohn's disease, but it is not my first drug of choice. I see it as valuable in attempting to wean Crohn's disease patients from steroids and when we have time to wait the three to four months that they require to take hold of the inflammatory process.

With careful monitoring of the blood counts, I am more comfortable with Imuran® and 6-MP in Crohn's disease than I am with steroids and have seen less complications with them. Intolerance to one is not improved by switching to the other. Florid pancreatitis (2 to 3 percent) is easily recognized. Milder forms of pancreatitis have been difficult to diagnose.

Cyclosporine A (CSA), a more specific immunosuppressive drug acting on IL-2 activated helper T-cells is yet to find a defined place in the treatment of IBD, but has the great advantage over Imuran® and 6-MP in that it acts much more quickly than they do. In ulcerative colitis, my personal experience has already demonstrated a place for Cyclosporine A in patients facing a relatively urgent total colectomy after a failure of seven to 10 days of IV steroids. In Crohn's disease, I see its place as being very similar to Imuran® or 6-MP when the clinical situation is urgent and the 12- to 16-week delay in their use cannot be enjoyed. But its efficacy in Crohn's disease is currently already been called into question. Oral absorption of CSA is erratic and renal toxicity is to be feared.

Candidate Drugs

Other drugs are waiting in the wings to have their place in the therapy of IBD evaluated. They include methotrexate, hydroxychloroquine and the Omega 3 fatty acids. *Methotrexate,* which has immunologic-modulating effects, has been introduced to the treatment of Crohn's disease in doses of 25 mg at weekly intervals. I have only very limited firsthand experience with this compound in a few patients with ulcerative colitis and await results of controlled trials.

I have participated in the controlled trial of hydroxychloroquine, the well-known antimalarial, which has suppressor T-cell activity, introduced by my colleague Lloyd Mayer, based on his demonstration of colonic epithelial cells failure to down-regulate helper T-cells in response to the intestinal content's antigenic load. It appears to be well tolerated with few side effects, but a dosage of 200 mg twice a day was effective in only some patients. A new trial of much larger doses (800 mg) is now being carried out.

With my colleagues, Asher Kornbluth and Peter Salomon, I have conducted an open trial study of the use of Omega 3 fatty acids as an anti-inflammatory food component in ulcerative colitis with moderate effectiveness in mild to moderate cases, as have others. Some controlled studies have also indicated a limited success with their immunosuppression of the leukotriene LTD 4. The large number of pills required to produce effective tissue levels limits their current effectiveness, but it is worth considering for a patient who would prefer a "food" to a "pill."

Inhibition of 5 Lipooygenase

Studies of the mediators of inflammation have shown that the leukotriene LTB 4 is a major chemotactic factor. I have already mentioned that fish oil contains eicosapentaenoic acid, which competes with arachidonic acid for 5 lipooxygenase.

Another approach to decreasing tissue production of LTB 4 is the use of a specific inhibitor of 5 lipooxygenase. One such selective 5 lipooxygenase inhibitor (Zileuton) has had some preliminary trials and seems superior to placebo in reducing symptoms, but conferred no advantage over Azulfidine®, nor added to Azulfidine's effectiveness.

At present there is ongoing a trial with a much more potent LTB 4 inhibitor, FK 506, which actually is a 5 lipooxygenase-activating protein inhibitor.

Selecting Drug Therapies

Faced with this cornucopia of available drugs, we must have some rationale for selecting the ones appropriate for the particular patient we are treating.

Drug Strategies for Ulcerative Colitis

We obviously need to know the degree of severity of the presenting syndrome—whether it is the first or one of several recurrent episodes and the anatomic extent of colonic involvement. I intend to present in this section the order in which I use our currently available medications starting with their use in the mild to moderately active degrees of inflammation.

PROCTITIS AND PROCTOSIGMOIDITIS

In these individuals with mild to moderate degrees of clinical and inflammatory activity and obviously well enough to be treated as outpatients, I favor 5-ASA and its variant forms as a first-line drug. If the proctitis is only a few centimeters in extent, the nightly use of 5-ASA (1 g) suppositories is a convenient and acceptable approach, but needs to be carried on for many weeks. For more extensive proctitis and proctosigmoiditis, topical rectal use of 5-ASA enemas is my most frequent approach using the 4-gram enemas of 5-ASA available commercially as the Rowasa® enemas. This method, whether it is the first attack or the recurrence of an acute attack in the case of a patient with a chronic history or frequent relapses, leads quite often to dramatic and prompt subsidence of longstanding bleeding, and which frequently occurs after only a few nights' use. The major problem I see in my consultative practice is that the prompt subsidence leads the patient and physician to a too prompt cessation of the drug by this route. I believe it should be continued for many weeks uninterruptedly until the sigmoidoscopy reveals an essentially grossly normal-looking mucosa; I do not insist on biopsy confirmation of healing, but I am persuaded by the gross appearance of clinical subsidence.

In the same context, if the patient is more uncomfortable, I reinforce the topical rectal route by administering 5-ASA by mouth as well. Unless there is a clear-cut history of allergic response to sulfonamides in the past or to any prior use of sulfasalazine, this compound either as the generic form or as Azulfidine® is my main reliance at this early stage of management. It is wise to start with small doses and gradually increase them to reach the projected level (total of 3 to 4 grams in four divided daily doses) in order to minimize the occurrence of headache, slight nausea, or a moderate decline in appetite. These symptoms often respond to small doses of antihistamines and will clear as tolerance develops with time. For those patients intolerant of sulfasalazine, the other available varieties of 5-ASA (mesalamine) obviously should be used in the well-known varieties: Asacol®, Dipentum®, Claversal®, Salofalk®, Pentasa®. With these, too, starting with small doses to reach the larger desired level is in order. These compounds are usually well tolerated in doses of 2.4 to 3.6 grams in divided doses of from six to nine tablets each; but, in our own double-blind study, as well as others, about 10 percent of patients on olsalazine experienced a rather marked increase in their diarrhea which precluded their use. Intolerance to both olsalazine (as the dimer Dipentum®) and the other variants of 5-ASA (mesalamine) can occur, but is rare.

In the period before weaning patients from the topical (rectal) use of 5-ASA, I almost invariably add 5-ASA by mouth and then slowly reduce the size and frequency of rectal instillation, advising their use on alternate nights, then lengthening the periods between episodes over the course of several weeks. In all, it may take up to four or more months to achieve complete remission of proctitis and of proctosigmoiditis by clinical and sigmoidoscopic criteria before I would consider the progressive return to maintenance therapy.

Some patients may experience some rectal irritation or burning sensation with the 5-ASA commonly used available preparation. This is possibly related to the preservatives used to maintain longer shelf-life. Some preparations with shorter shelf-life without the preservatives can be obtained. Sometimes the instillation of a small amount of hydrocortisone in the form of Cortifoam® or Proctofoam® some time in advance of the 5-ASA preparation may be useful.

But proctitis is a stubborn disease, like eczema, and hangs on presenting difficulties for the patient's discomfiture and to the physician's chagrin. My first step at this point is to add the use of Cortifoam® as a rectal instillation in the morning for those of us who must go to work and continuing the additional 5-ASA enema at night. For those that have the time and for all on weekends, I advise administering the rectal instillations of the 4-gram 5-ASA enemas twice a day—in the morning, as well as at bedtime.

If the rectal bleeding, tenesmus, and frequent bowel movements of proctitis and proctosigmoiditis persist, patients, despite their initial reluctance, are willing to escalate their therapy to rectal steroids, usually hydrocortisone, 60 mg per enema. I insist on their continuing oral medication in the form of sulfasalazine or mesalamine in anticipating a remission. Unfortunately, some patients, even with this program, must be switched to oral steroids. Experience has convinced me that it is wise to start with a reasonably large dose to induce remission rather than with a small dose and then be forced to add further increments. It may on occasion require 45 and even 60 mg of prednisone to accomplish significant reductions in symptoms and bleeding. Once clinical remission is achieved, measured by clinical improvement and by sigmoidoscopic appearances, then the process of slow weaning should be instituted while the patient remains under the "umbrella" of oral or rectal 5-ASA. It is wise to warn the individual of the likelihood of significant absorption of rectal steroids when these are still being employed. When I use reasonably large doses of oral steroids, I tend to stop their rectal use. In the past, on rare occasions, I resorted to IV steroids to reduce a remission, but not in recent years. I can even foresee the possibility one might need to institute azathioprine or 6-mercaptopurine in proctosigmoiditis in an effort to wean patients from oral steroids who relapse when the steroids are reduced below a critical level. The exorbitant costs of cyclosporine enemas have so far precluded my attempted use of them in ulcerative colitis.

When reducing the oral steroids from initially high doses of prednisone, I come down by 5-mg decrements. When I approach the 15- or 10-mg level, I then use 1-mg decrements. This weary catalogue of progressive maneuvers demonstrates that the treatment of proctitis and proctosigmoiditis can be tedious and protracted for this stubborn disease. I should also add that I have not found a place for antibiotics, including metronidazole, in ulcerative colitis in most instances.

LEFT-SIDED AND UNIVERSAL ULCERATIVE COLITIS

These varieties of ulcerative colitis can be of mild to moderate severity and can be treated with outpatient management. In the mildest forms, a short trial of oral 5-ASA, either sulfasalazine or mesalamine, is worthwhile. But in those with systemic symptoms, such as anorexia, weight loss, low-grade fever, fatigue, along with the customary loose stools and rectal bleeding, oral steroids are my first choice. In some with considerable rectal discomfort, I add 5-ASA enemas and reinforce the dose of 5-ASA by the oral route. For those who are sick but are clearly not severely ill, an alternative to hospitalization and IV steroids is the outpatient use of intramuscular adrenal corticotropin (ACTH) in gel form on a daily basis or perhaps five times a week at a dose level of between 25 to 40 units a day for several weeks. This may stabilize the illness and avoid hospitalization, but stubborn instances of protracted mild to moderate pancolitis may require seven to 10 days of IV steroids. (I discuss the details of this in the next section "Severe Ulcerative Colitis.") When patients are being weaned from the oral or IV steroids or the intramuscular ACTH, I continue them on good doses of 5-ASA (3 to 4 grams) during this period. In the absence of suppurative complications, I do not see the value of antibiotics.

SEVERE ULCERATIVE COLITIS

Whatever one's criteria are, severely ill patients with ulcerative colitis should be hospitalized. On the basis of our own double-blind controlled randomized trial (ACTH versus hydrocortisone) and our review of the published material of well-defined open trials, I have come to formulate this kind of program. These very sick individuals should receive either hydrocortisone by vein, 100 mg during an eight-hour infusion, or 40 units of ACTH in the same period of time (three times around the clock) for 24 hours a day for at least seven and possibly 10 days. Since I feel one should aim at maintaining a high circulating level of steroids rather than aiming at maximal physiological levels, I institute a continuous IV administration of the steroids rather than three peak pushes by soluset. At this stage, if the patient's appetite is poor, I see no place for 5-ASA but would add this to the drug program when the patient is being switched over to oral steroids or weaned from the ACTH.

When the patient has not been on oral steroids for the month preceding the hospital admission, we are convinced ACTH is the more effective drug. If they have been on oral steroids, IV hydrocortisone is the drug of choice. Even former skeptics have come to accept this approach. All experienced physicians have known for a long time that there is a place for ACTH in the treatment of severe ulcerative colitis. While others have added IV antibiotics, steroids, or 5-ASA by mouth and/or by rectum in severe ulcerative colitis patients, I rely primarily on the IV steroids (hydrocortisone or ACTH), reserving antibiotics for suppurative complications, including toxic dilation, although recently, antibiotics have been advocated by careful observers.

In these severely ill patients, the important question is how long one should continue these high doses of IV therapy. British researchers have drawn an arbitrary line at five days with surgical intervention if the patient has not responded properly. In our own studies, in which some patients required nine days, the mean was $7^{1}/_{2}$ days and some required even longer periods, but, I must confess that, on occasion, I have continued the steroid course for 14 days. The therapeutic path has now been opened with another escape hatch at this junction. Recent experience at Mount Sinai in an open and now a controlled randomized trial of cyclosporine for seven days has persuaded me to employ this drug as a last ditch stand with considerable success.

Risk factors of the standard IV regimen are few. Steroid therapy does not worsen the outlook of these sick patients. It does not increase the risk of perforation, although it makes the diagnosis more difficult by masking the clinical signs of fever, tachycardia, and the physical signs of abdominal wall tenderness or rigidity. Daily percussion of the abdomen for liver dullness is a sine qua non and frequent flat plates of the abdomen have lessened the risk of missing this complication. I have personally not seen in more than a 100 patients the adrenal cortical hemorrhages recently reported in a few cases. It may be prudent to obtain an abdominal sonogram (or even a CT scan) of the adrenal area before starting this therapy and have a baseline for any changes in the size of the adrenals. The reported patients have all recovered without significant morbidity. In view of the general concern among lay persons, as well as physicians, regarding the ulcerogenic potentiality of large IV doses of steroids, the question arises regularly regarding the prophylactic use of antacids and the use of histamine$_2$ blockers of gastric acid secretion. Generally for the patients referred to me or transferred to this hospital, the majority of practitioners apparently use H2 blockers routinely. Since there are no definitive studies on this point in these ulcerative colitis patients who do not have an excess of peptic ulcers I do not routinely use antacids or H2 blockers, except in those with peptic symptoms.

Once the patient has improved on IV therapy, the next step that must be faced is the transition to oral medication to facilitate hospital discharge. For those on hydrocortisone, or its equivalent, the choice is simple—one must switch to a comparable dose of oral steroids and begin the slow weaning process. Here I almost invariably add the oral tolerated form of 5-ASA to ensure the continuation of the remission. With the response to IV ACTH, the choice is a bit wider. One can either rely on the remission and continue without any exogenous cortisone or, fearing a relapse, add a significant dose of oral steroids plus 5-ASA in any tolerated form. I do not know as yet which route is more effective, so I continue to alternate them, but, bear in mind that the number of patients entering the hospital with severe ulcerative colitis while not on current steroid therapy are relatively few. In our published experience concerning IV ACTH and IV hydrocortisone, it took six years to obtain the needed 66 patients (33 for each arm)!

THERAPEUTIC STRATEGIES DURING THE WEANING PERIOD

Most patients treated with steroids, no matter what their previous dependency, once they go into remission, can usually be weaned from steroids over varying periods of time. The major problem is not the development of withdrawal symptoms (anorexia, abdominal distress, "gooseflesh," yawning, and diffuse joint pains), but the appearance of a relapse during the slow weaning period. In some instances, the problem is easily solved. If the indications for colectomy, which were narrowly avoided by the most recent intensive medical IV steroid or cyclosporine program are still present, the patient and the physician will be convinced that surgery is the logical sequence. In others, especially patients with left-sided or pancolitis, who are relapsing, but are not sick enough at this juncture to warrant colectomy, one should consider other alternatives: first, a return to the last dose of oral steroids on which they were comfortable, and/or second, continuation of 5-ASA by mouth, or third, institution of immunosuppressant therapy with either azathioprine or 6-MP, providing one has time to maneuver as both drugs require at least eight and possibly more weeks before they take hold.

One should add at this point that a clinical difficulty arises in a few patients without prior peripheral joint pains when they are started on IV steroid therapy. These few patients develop severe peripheral arthralgias that subside as the dosage of steroids is lowered or the patient is weaned from the steroids. Detailed studies of these few patients with my rheumatological colleagues has led us to believe that these patients are not having a reactive arthritis to their underlying disorder, but really are having a reaction to the steroids themselves. For the moment we are labeling this syndrome "Pseudorheumatism."

MAINTENANCE DRUG THERAPY IN ULCERATIVE COLITIS

Patients with ulcerative colitis have an inherent tendency to relapse, the likelihood of which I have indicated in the section on the natural history of IBD (Chapter 14). Thus, prevention of such relapses is a prime requisite for any rational, long-term, therapeutic plan.

I should express my underlying bias regarding maintenance therapy in IBD; since all our medications are not directed toward the putative agent(s) of the disease, I believe that prophylactic therapies are essentially an attempt to suppress mini-relapses. It is also rational to continue a medication on a program that has induced a remission in a given patient, provided, of course, that the risks of the long-continued use of that medication do not have outweighing risks and side effects.

The last proviso, it seems to me, rules out the use of steroids in any reasonably prolonged maintenance program. Of the disastrous side effects, aseptic bone necrosis has been the most alarming in my personal experience. In addition, the immunosuppressive effects of steroids predisposing patients to bacterial, viral (especially cytomegalic virus), and fungal infections, provide further reasons to avoid long-term steroid maintenance therapy in ulcerative colitis. While alternate-day therapy reduced the risks of adrenal insufficiency in children, its long-term use in my hands has not been an effective, safe approach, except in a small number of highly selected patients who have a disease which can be "cured" by colectomy. This does not mean that we all have not seen patients who continued to take small doses of prednisone over long periods of time without major catastrophes. They are usually individuals who have a long personal history of ulcerative colitis and a sense that they "know what they are doing." Reluctant to have their practices reviewed, they are also the ones who avoid long-term cancer surveillance. Poorly absorbed or "nonabsorbable" steroids, especially by topical rectal approach now under study may become available, as in the form of budesonide, and may allow us to reconsider their long-term use, but certainly none at present have been documented to be safe over long periods of time.

MAINTENANCE WITH SULFASALAZINE AND
OTHER 5-ASA VARIANTS

Present experience, as well as the published maintenance trials, with one exception, have convinced me that sulfasalazine (Azulfidine®) and 5-ASA are effective and safe medications that are clearly superior to placebo therapy in maintaining ulcerative colitis patients in remission. In my patients, 1.5 to 2 grams of sulfasalazine given in divided doses, has been most generally used. A few individuals have found that they will do better with some larger dose, perhaps in the range of 3 to 4 grams.

As might be expected, 5-ASA (mesalamine) in several forms for delivery and release in the colon (Dipentum®, as a dimmer, olsalazine) and delayed release forms of mesalamine (Asacol®) or balsalazide (Colazide®, 5-ASA linked to 4-aminobenzoge-B-alanine) are also quite effective in prolonging relapse-free periods in ulcerative colitis once the patients are in remission. Obviously it cannot be expected to have any advantage over the older sulfasalazine form except its tolerance by individuals sensitive to the latter drug. Topical use of 5-ASA enemas can also attain remission in proctitis and proctosigmoiditis and left-sided ulcerative colitis, but have a distinct nuisance value in patients who are reluctant to continue the longterm instillation indefinitely. If patients have responded to 5-ASA enemas only to relapse on weaning or the substitution of mesalamine by mouth, I have at times fallen back on the long continued use of rectal instillations for periods of up to six months, attempting to get the patient onto a longer, every-other-night, regimen.

What about maintenance in ulcerative colitis with the immunosuppressant drugs? If a patient has gone into remission with azathioprine or 6-mercaptopurine in the course of ulcerative colitis or is finally weaned from steroids by their use, should they be maintained on this medicine? My colleagues Daniel Present and Burton Korelitz have convinced me of the relative safety of 6-MP and my own experience with azathioprine is similar, but I still have some reservations for the indefinite use of both. I am reluctant to continue these drugs after six months and certainly after 12 months, but patients in remission are equally reluctant to give up the drug that they believe has saved them from a colectomy and ileostomy. I am even more reluctant to continue the long-term maintenance therapy of ulcerative colitis with cyclosporine A in those few patients whom I have treated with that drug who have gone into remission and escaped colectomy. Somewhere in the sixth- to twelfth-month period of maintenance, I attempt to wean them to azathioprine or 6-MP.

Drug Strategies for Crohn's Disease

Our aim of therapy in Crohn's disease, of course, is to induce a remission of clinical symptoms, promote tissue healing, and maintain that remission, yet we know at the same time that the susceptibility to the disease is lifelong, the etiology unknown, and the tendency to relapse inherent in the disorder. Our solace is the fact that patients in a remission, once achieved, have a tendency to continue in that remission for a long period of time, even up to one year.

Let me make my biases clear before I outline my approach to treating patients with Crohn's disease. I do not separate the therapy of small bowel disease, especially from colitis or ileocolitis. Crohn's disease is Crohn's disease is Crohn's disease, even though I attempt to distinguish the more aggressive from the indolent variety. Further, I expect that any program that induces a remission should be used to maintain that remission except for the long-term use of steroids. Further, my prejudice against steroids is based in part on the appearance of very little visible effect of these drugs on the histopathology of Crohn's disease and their increasing the risk of suppurative complications of the disorder.

In considering the order of my use of current drugs in Crohn's disease, I turn first to the patient with mild to moderate disease whom it is reasonable to attempt to treat on an ambulatory basis. Sulfasalazine is my first choice in all forms of CD in the context of mild to moderate clinical disease activity. I begin with small doses of these compounds, and gradually increase them in time to the point of tolerance to avoid early nausea and/or vomiting especially, reaching up to 3 to 4 grams in divided doses and rarely go beyond 4 grams. I use the coated (Entab®) form of Azulfidine® only in those with more marked upper gastrointestinal complaints. Sulfasalazine of course requires an obligatory dose of at least 1 mg of folate or folic acid at the least. With the availability of the newer 5-ASA variants as mesalamine or olsalazine, I no longer attempt to desensitize individuals intolerant to sulfasalazine. My initial dosage would be 2.4 to 3.6 grams in divided doses. I rarely go to 4.8 grams. Since I do not know whether 5-ASA interferes with folic absorption or not,

I usually ask the patient to take at least 1 mg daily. Since it was easier up till recently in the United States to obtain olsalazine rather than the other variants, I attempted to start with that compound first, but I am now constrained by the fact that in our published studies on patients from Mount Sinai Hospital, 10 percent of our patients on this drug have significant diarrhea. If patients are already on 5-ASA in that form, I continue them while warning them about the possibility of diarrhea rather than switching to Asacol® or Pentasa®.

If rectal discomfort is important or left-sided involvement with Crohn's disease has been demonstrated, I may reinforce the oral sulfasalazine or oral mesalamine or olsalazine with topical rectal instillations of 5-ASA in the form of 4-gram enemas. Anal Crohn's of limited extent may profit also from 0.5- to 1-gram mesalamine suppositories nightly.

I am aware of the belief held by some clinicians that sulfasalazine has little value in CD limited to the ileum or small bowel, but my own experience convinces me that the drug is of value. To the theoretical objection that sulfasalazine fails because of lack of bacterial splitting of the compound in the small bowel, I believe that small bowel bacterial overgrowth of colonic organisms occurs frequently in Crohn's disease when there is any narrowing of the lumen, fistulization of the colon, or anastomosis of the colon, so I continue to use it as the first drug. If the use of sulfasalazine is insufficient to reduce a remission in the mild to moderately sick patients in a trial of at least four weeks, judging from the history of patients referred to me, it seems almost the inveterate practice of gastroenterologists to have begun with steroids or to use steroid therapy jointly with sulfasalazine right from the beginning of therapy no matter what the degree of severity of the involvement being treated. For the reasons I have already alluded to above, this is not my practice. I turn next to antibiotics.

ANTIBIOTICS INCLUDING METRONIDAZOLE

There exists in Crohn's disease what might be called an "underground" of therapy, widely practiced but rarely written about, which is the use of antibiotics directed against intestinal inhabitants.

Two comments seem in order. Our predecessors smart enough to recognize these forms of IBD thought there was a place for antibiotics and my own predecessors relied heavily on the nonabsorbable sulfonamides, especially sulfathaladine. Some patients have continued to rely on this group and have even stockpiled the last of these drugs since they are no longer commercially available in the United States. Further, many competent and careful observers of the present do use the drugs of the ampicillin and tetracycline variety in patients with Crohn's disease in the absence of frank complications, such as well-established abscess or draining perirectal fistula but few report results. There has been lately an increasing tendency to use ciprofloxacin in this context. It was unfortunate that the mechanism that existed for controlled trials of antibiotics in Crohn's disease, which was developed for the National Study, was dismantled before this group of drugs was investigated. In the past, I relied on alternating courses of antibiotics directed toward the aerobic and especially the anaerobic organisms of the lower gastrointestinal tract.

Those advocating antibiotics as primary therapy in Crohn's disease may find some rationale in studies describing abnormal intestinal flora found in many Crohn's disease patients, as well as those actually demonstrating their presence in the bowel wall and mesenteric lymph nodes.

Metronidazole as drug therapy in Crohn's disease. Metronidazole, which has considerable antimicrobial activity against intestinal anaerobes, deserves separate discussion in the context of the use of antibiotics since it also has some immunomodulating effects. I have used it, since its introduction in 1975 by Ursing and Kamme, not only for perirectal disease with fistula formation, but also for ileocolonic Crohn's disease above the rectum and perineum usually along with other antimicrobial agents, such as ampicillin and Cipro®. I have found that most patients tolerate 1 gram daily in two or four divided doses. They accept the metallic taste, the injunction against alcohol, and the possibility of mild abdominal cramps. The disturbing side effects of the neuritic variety, including numbness of the extremities, usually the feet but not always only, usually clear on reduction of the dosage or stopping the drug entirely, but this may take a long time to resolve.

I discuss drug therapy and the complications of Crohn's disease in the perineum elsewhere, but I continue to be impressed by metronidazole's usefulness in perineal and rectal abscess formation, and in fistulization of Crohn's disease, but I sense that it may require long periods of time for healing of these situations.

STEROID THERAPY

If sulfasalazine or mesalamine and its usual variants plus concomitant antibiotics, including metronidazole, fail to induce a remission and improve patients, I turn, although reluctantly, to steroids. Corticosteroids make patients with Crohn's disease feel better, suppress fever, improve appetite, lead to weight gain, and shut off intestinal mucosal protein leak. Their efficacy in this context has been clearly demonstrated in short-term studies in patients well enough to be managed on an outpatient basis. Judging from a referral practice, almost all patients with Crohn's, sick or not, receive steroid therapy these days. That this drug alters the long-term course is certainly not known, although widely assumed. Indeed, we know little about the effects on the histopathology of the disease; they certainly have helped the extraintestinal or colitic manifestations. Their growth-retarding effects in children are well known, but must be measured against the beneficial suppressing effects on the inflammatory component of the disease. While side effects are reduced by giving corticosteroids on alternate day programs, most patients do not feel as well on the skipped day as on the day when the drug is taken.

As in ulcerative colitis, the disappearance of the sense of well-being and the reappearance of these manifestations as the drug levels are reduced, leads unfortunately to prolonged courses of corticosteroids and their attendant hazardous complications—cataracts, aseptic bone necrosis, peptic ulceration, and mood alterations, along with frank psychotic episodes. My current stance is that steroids are useful in short courses for improving the clinical status of the patient. I do not think they are deleterious in the short run and, unlike some skilled physicians, I fear their long-term use. I share my patients' anxiety about this class of drugs. The addition of sulfasalazine to steroids does not seem to improve the therapeutic advantage, although this combination is widely used. As have others, I have found that, if steroids are to be used, it is wiser to begin with a significant large dose rather than with small doses which then have to be increased step-wise. It is more effective to use a larger dose at once and then decrease that dose.

I use corticosteroids in the treatment of both small and large bowel Crohn's disease, although with reluctance. Care should be taken to avoid initiating steroid therapy in the presence of infected complications, such as perianal or intra-abdominal abscesses, in which circumstances I favor antibiotic therapy. Once the infection is controlled, corticosteroids may be added if the patient does not respond to 5-ASA in any of its forms. I begin with 45 to 60 mg of prednisone daily administered in divided doses and continue this course for approximately 10 to 14 days, at which time I begin to taper off 5 mg every seven to 10 days. This reduction should be in the evening doses to diminish the side effects of insomnia. The dose of prednisone must be raised or the rate of reduction slowed if relapse of disease activity follows the tapering process.

WEANING FROM STEROIDS

Once in remission, it is obviously desirable to withdraw the patient from steroids. Earlier clinical trials have suggested that at least one-third of patients develop symptoms within one year as the dose of prednisone is reduced. In patients with quiescent disease, the U.S. National Cooperative Crohn's Disease Study found that the withdrawal of corticosteroids was not associated with a greater relapse than was the continuation. However, in the setting of acute disease, prednisone withdrawal is associated with clear deterioration more often occurred with ongoing prednisone therapy. Therefore, when withdrawal of prednisone is attempted in these patients in clinical remission after the maximal therapeutic dose is achieved, success is usually accomplished. Once in remission, we aim to withdraw the patients from steroids. Patients are reluctant to give up their sense of well-being and fear inevitable relapse based on their previous experience with Crohn's disease. It is here that in weaning the patients we can get into trouble. I can see no place for maintenance steroids in Crohn's disease, although I am well aware that some clinicians do maintain their patients on small doses of prednisone, perhaps 5 to 10 mg daily, and I know that some of my own patients are surreptitiously doing the same.

Drug Therapy in Patients Severely Ill with Crohn's Disease

In the very sick patient who does not tolerate the oral therapy or is not responding to it, parenteral corticosteroids, including hydrocortisone or adrenal corticotropin (ACTH), are my fallback position. In the absence of controlled trials, I am convinced by our controlled ulcerative colitis trials of ACTH versus hydrocortisone to use ACTH in those who had not been treated with corticosteroids for 30 or more days (a very small group these days), and to use hydrocortisone for those who had been on oral steroids (the majority). I administer 100 mg of hydrocortisone or 40 units of ACTH in 5 percent dextrose and water intravenously by constant drip. These medications are infused over each eight-hour period to deliver a total daily dose of 300 mg of hydrocortisone or 120 units of ACTH for seven to 14 days. Intravenous therapy is discontinued once the optimal benefit is achieved and oral prednisone 45 to 60 mg a day is begun again.

Problems with Weaning from Steroids

To try to ensure the maintenance of remission during the weaning period, I add or continue the patient on the variants on 5-ASA he or she tolerated before and use maximal doses and often add or continue metronidazole. While the majority of decremental substeps are lowered every seven to 10 days, as we near the 10- to 15-mg critical level, I slow down the weaning to 1-mg steps and try not to jump back a step when the very first objective symptoms appear and try to persuade the patient to tolerate these "bumps" on the road.

It is not rare to find that the weaning process is unsuccessful and reincidence of disease surfaces, tempting us to try again suppression by an increase in steroid dosage, only then to find that symptoms reoccur as one resumes the slow, tedious weaning process. Here at this clinical point in decision making, we must decide whether the mechanical aspects require mechanical treatment, that is surgical intervention, if not we then have the option of immunosuppressant drugs; this is a convenient time to discuss their place in my treatment of Crohn's disease.

Immunosuppressant Drugs and Crohn's Disease

Probably the major change in my own thinking about drug therapy in Crohn's disease in the last eight years from the earlier edition of this book has been my acceptance and reorientation regarding the place of the immunosuppressants, azathioprine (Imuran®) and 6-mercaptopurine (6-MP) (Purinethol®). My colleagues and former students, Present and Korelitz, have convinced me regarding the safety of these compounds which I consider interchangeable, since Imuran® is converted to 6-MP in the body, and I have not found that an individual intolerant of one tolerates the other any better. The short-term use of these drugs, if monitored by frequent blood counts, has not presented important problems. Pancreatitis has been more troublesome, reaching in my own experience between 5 and 7.5 percent. The long-term studies reported by these workers have emphasized their safety and their very limited carcinogenic potential. Very long-term studies still remain uncertain and this underlies my reluctance to maintain patients on these drugs indefinitely.

From the tenor of my discussion so far, it is obvious that Imuran® and 6-MP are not my choice of first-line and second-line drugs in the mild to moderately ill patient and, in the severely ill patient, the eight- to 12-week delay in being effective reduces their usefulness. My prime use of this group of immunosuppressants is in the period of slow weaning from steroids, especially when the first attempts have failed and the patients have been reluctant to return to the original larger doses of steroids. Here I am more comfortable with azathioprine simply because of my longer experience, which in doses ranging in the 50 to 150 mg daily has been most helpful in allowing me to get the patient off steroids and to continue on 5-ASA compounds. I shall discuss their maintenance in the next section and in Chapter 24 the special problem of fistulization.

What about the sicker patient who cannot tolerate the longer latent period of 6-MP and Imuran®? Further, is there a place in others for cyclosporine A? Cyclosporine A, an immunosuppressant drug which has revolutionized transplant surgery, is a more focused immunosuppressant than Imuran® or 6-MP acting on helper T4 cells and with a more rapid onset of action. I have experience with some patients where an attempt with a short trial of this drug was important, mainly in Crohn's colitis patients facing colectomy after failure to respond to IV steroids or IV ACTH. Here a trial of IV cyclosporine, appropriately monitored for kidney function for seven- to 10-day periods, has, in the few patients tried, had a favorable effect for the time being. If this attempt is successful and the patient is weaned from steroids, I substitute azathioprine or 6-MP after about six months.

For the mild to moderately ill patients with Crohn's disease, I have looked on cyclosporine as a more rapid Imuran® or 6-MP and interpret the published literature to mean, that while it can produce a more rapid remission, it is followed by a more rapid relapse. Accordingly, I do not routinely advise its use in this group of patients.

Maintenance Therapy in Crohn's Disease

While most published studies indicate that when patients with Crohn's disease go into remission or improve from whatever therapy they are on, they have a tendency to maintain that remission for periods of time, even up to more than a year. Yet we are all aware of the inherent tendency to relapse with fairly prompt reappearance of microscopic and gross pathologic changes on endoscopic study and biopsy. Thus the need for effective maintenance therapy is of foremost importance. It has always been assumed that almost all studies, until recently, including our own meta-analysis of maintenance therapy in Crohn's disease, have not demonstrated the effectiveness of any simple drug or combination of drugs. Maintenance therapy I believe is the suppression of mini-episodes and it is difficult to understand why any drug capable of improving the acute situation does not continue to be effective in the maintenance phase. The more recent reports from the Canadian and European studies on mesalamine in Crohn's disease of patients in remission following surgical resection demonstrate a significant beneficial effect in reducing the incidence of relapse. These reports, as well as the incidence of patients with prompt relapse after surgery, have encouraged me to attempt maintenance therapy with 5-ASA simultaneously while weaning them from steroids and promptly after operation in all cases. I use whichever variety of 5-ASA orally the individual patient has been shown to tolerate previously. My present stance is to attempt to continue this maintenance program indefinitely. If it has been necessary to resort to immunosuppressants, such as azathioprine and/or 6-mercaptopurine, to wean the patients from steroids, I am committed to a longer maintenance program with these compounds, but how long remains an unanswered question. I am reluctant to continue the patient for more than a year, while those individuals with previous clinical courses that were stormy are reluctant to abandon this compound which has stopped or prevented the continuation of their disease, including repeated surgical resections. A few determined individuals in my practice have remained on Imuran® therapy for more than six years with generally low toxicity and a continuing favorable course, but not always uniformly so.

As in ulcerative colitis, in the patients treated with cyclosporine for Crohn's disease I have maintained them on cyclosporine for six months and then continued to rely on 5-ASA alone; if I have switched one or two to Imuran® or 6-MP therapy, I ultimately pressure them after six months to rely on 5-ASA alone too.

Other Drugs for Maintenance Therapy in Crohn's Disease

What is the role of *metronidazole* in maintenance therapy in Crohn's disease? I must confess considerable ambivalence regarding this medication with its long-term side effects, especially the neuritic ones. Patients with perineal disease, to be further discussed in Chapter 24 in the section on fistulous complications, often remain on and require long-term maintenance with metronidazole. But I resort to the use of the surgical Parks procedure rather than maintaining these patients on long-term indefinite metronidazole. I am not convinced of its usefulness as yet in the prevention of relapse in Crohn's disease above the rectum and perineum.

Methotrexate, which has antimetabolic, anti-inflammatory, and folic acid antagonistic properties, is enjoying a wide vogue at present in diseases such as primary biliary cirrhosis and sclerosing cholangitis and has been used in an open study of a moderate number of patients with ulcerative colitis and Crohn's disease with some interesting and encouraging results. I have had limited firsthand experience with it in either disease, and I am looking forward to further studies, including larger open studies. The need to prohibit conception in women and the as yet unavailable information on its long-term toxicity and pathology emphasize the need for caution, but it does give some hope for some time in the future. Although I have an ongoing interest in treatment of hydroxychloroquine (Plaquenil®) and Omega 3 fatty acids from fish oil in inflammatory bowel disease, I have used these compounds only in ulcerative colitis and have no information or experience regarding their effectiveness in Crohn's disease.

16

Supportive Supplemental Therapy

Nutritional Considerations

If we cannot cure these patients, as we must reluctantly confess, what can we do to improve the quality of their lives?

If anemia contributes to the fatigue and weakness so many sufferers of IBD complain of, then correction of anemia due to ulcerative bleeding, B12 deficiency due to ileal dysfunction, and folate malabsorption due to the interference with intestinal absorption by sulfasalazine need correction. Unfortunately, all too often the anemia is due to the chronic inflammatory process with inhibition of red cell maturation in the bone marrow; erythropoietin's role is unclear.

When patients and their families realize that they must deal with the ongoing problems of inflammatory bowel disease, they seem preoccupied in my experience with two questions: What should we eat? and How important is psychological stress?

Around no other aspect of medical therapy of these disorders is there so much patient and family discontent as that which surrounds the question of what diet the sufferer is to eat. Our society places so much emphasis on food; the cultural climate is so concerned with the "natural." Legend and folklore plus deep psychic concerns over what we ingest re-enforce all our instincts that food must play some role in affecting the health of our intestinal tract. No wonder patients are troubled over whether their past dietary faults have led to their present difficulties; and even more important they want to know what they should eat to get better and better faster. No wonder they are disappointed in the little we have to tell them and no wonder they consult a wide variety of nutritionists even when they follow our drug therapy advice. I see no way at present of meeting their expectations and current frustrations except by devoting a good bit of time to a discussion of the role of nutrition in this disease. This means for most people reviewing with them what we consider the elements of a well-balanced, satisfactory, "normal" diet. Requirement for a desirable high level of protein intake needs to be emphasized with some accent on the problem of protein leaking from the inflamed mucosa. The basic action of the accessory food factors need to be briefly sketched. The roles of vitamins and trace minerals need not be left to the naturopaths' didactic teaching. The physician must also admit with candor how limited current information is in IBD in contrast to the striking examples of lactose intolerance or gluten enteropathy.

Diet and Nutritional Therapy

From the etiologic point, it is far from clear that any deficiency or excess of nutrients plays any role as an environmental factor in ulcerative colitis. I do not know what to do with the fact that Crohn's patients apparently eat more sugar and less fat than their controls in a great many case control studies. Patients and their families need to be reassured that the subject's prior eating habits played very little or no part in inducing the disease or setting the stage for its appearance. I have the impression that our patients with Crohn's disease, in contrast to ulcerative colitis, have a greater proportion of sufferers from atopy or atypical allergic disorders—hay fever, eczema, hives, asthma—than their controls. It is when we turn to the management of the individual patient that the flood of questions regarding diet erupts.

Lactose intolerance has been so present in the lay literature that this area surfaces quite quickly when dietary factors are discussed with patients. Some lactose intolerance does exist in IBD, but it is far from universal. In my experience, by the time patients come to see me, they all have usually discovered their intolerance for milk, cheese, ice cream, and butter. I have found that patients' experience with these foods more helpful then the standard lactose tolerance test, which I rarely do. If there is any uncertainty in this context, I suggest a two-week trial of a low-lactose diet, interdicting the major sources of lactose, and do not encourage the rigorous trials that would involve the reading of labels of all foods eaten. I do not veto a trial of the available lactase additive (Lactaid®) to milk for those who feel its avoidance a great hardship, but neither do I encourage it.

Since folate concentration in the jejunum and serum levels are increased by bacterial overgrowth, most patients have normal blood levels. Dietary folate needs to be supplemented, however, in patients taking sulfasalazine. Vitamin B12 in appropriate doses is needed when the ileum is present and diseased or absent due to resection. Much has been made of zinc; difficulties in florid skin forms do manifest themselves occasionally in patients with short bowel syndrome, after long malnutrition, or in those receiving TPN. Calcium and Vitamin D supplements may reduce the osteopenic complications of prolonged steroid therapy.

All physicians treating IBD soon realize that their patients are assaulted by an environment of nutritional advice from friends, relatives, daily newspapers, and all the media, to say nothing of the popular how-to books by nutritionists and self-appointed dietary propagandists, so we need to formulate our own approach and inform our patients early on as to what we know and especially what we don't know.

There are many reasons why patients with chronic ulcerative colitis and Crohn's disease may be malnourished. First, decreased intake of food because of nausea, vomiting, early loss of appetite, and diarrhea. Second, malabsorption because of small bowel disease, bile salt catharsis, and malabsorption following ileal disease with bacterial overgrowth along with B12 deficiency. Third, nutritive loss from bleeding iron, protein loss leaking from the small or large bowel, electrolyte loss from diarrhea involving trace metals—zinc or selenium. Fourth, increased requirements for the catabolic state of fever, infection, and inflammation, as well as steroid therapy. Fifth, effects of our drugs—steroids on calcium and bone metabolism, sulfathalazine on folate absorption, bile binders such as cholestyramine—and these must be addressed if at all possible before our patients suffer from weight loss, hypoalbuminemia, anemia, low serum Vitamin B12 and folate, and a few with low magnesium, some with low potassium blood levels, and rarely from zinc deficiency.

Despite this long catalogue of possible and probable causes of malnutrition, most patients with mild to moderate IBD, whether of the ulcerative colitis or the Crohn's variety, seen early on in the course of the illness are not seriously nutritionally compromised. The drug measures already discussed with their reduction in the inflammatory process are the important therapeutic maneuvers.

In the presence of the narrowing of the gut from edema, stricture, inflammation, and/or surgical scarring, it is wise to reduce the nonabsorbable fiber in the diet; and if the patient is suffering from diarrhea, it may make sense to restrict dietary fiber and thus reduce the work of the colon.

Nutritional Supplements: Enteral and Parenteral

In adults, supplementation of ordinary foods by the nonresidue oral liquid formula diets, such as Sustagen®, Susta-Cal®, or Ensure®, may be helpful in those with the capricious appetites of sickness or the drug-induced anorexia of sulfasalazine. In adults I have never found the need to have prolonged periods of nocturnal nasogastric drip of these dietary supplements (Flexi-Cal® or Vivonex®) probably because the problems of retarded growth and sexual maturity marked with young patients are not present. Their use in children, however, is quite remarkable as recorded by others.

Along with the hypothesis suggesting that a diet high in sugar and/or low in fiber may be involved in the etiology of Crohn's disease, has been the proliferation of reports in the literature regarding the value of treating this disease with a fiber-enriched and/or sugar-free diet. Fiber rich and unrefined carbohydrate poor diets, combined with conventional medical therapy have had a favorable effect on the course of the disease measured by the need for hospitalization and surgery without leading to intestinal obstruction. Although several studies emphasize the therapeutic benefits of sugar-free diets, these studies are inadequately controlled with only a trivial statistical significant difference in patients compared with controls and provide unconvincing evidence on a clinical and biological scale to support their therapeutic role in the therapy of Crohn's disease.

Total Parenteral Nutrition

The feasibility of total parenteral nutrition (TPN), now universally available, has resulted in its use as an adjunctive therapy, especially in preparation of depleted patients with low serum albumin for surgical extirpation of the colon, although removal of the colon is probably the safest and fastest way to stop this protein leakage.

The interest in TPN as a major form of therapy extensive in Crohn's disease, especially for patients with cutaneous fistulas, has raised the problem of the place of TPN in patients with ulcerative colitis, since most intensive programs for treating a toxic ulcerative colitis bowel stress bowel rest. In some controlled studies, no difference between treated groups and control groups as judged by their clinical need for colectomy resulted from the use of TPN. In more recent studies of carefully controlled trials with patients receiving TPN along with other medical therapies, usually steroids, TPN did not improve the outcome. And finally invariably one may question whether TPN effectively rests the bowel in any significant quantitative way. Readers of a previous generation may recall in this context the past flurry of research of the use of "medical ileostomy" by long-tube intubation and draining of the ileal contents in patients with ulcerative colitis in the vain hope of providing bowel rest for the colon. In severely ill patients with ulcerative colitis, I have no objection to but see little advantage in TPN in the five to 10 days of intensive IV steroid therapy and have no strong feelings that the use of TPN at this stage of illness will significantly improve the chances of patients escaping colectomy or having a better postoperative course. In these depleted and partially immunocompromised individuals, I am concerned about the risk of sepsis and fungal infection arising from the use of the central line so at present I use TPN rarely in patients ill with ulcerative colitis and I find no convincing evidence on others' experiences revealed in the literature. My own experience leads me very much to the same point of view regarding the long-term use of TPN in Crohn's disease, except in those patients suffering from short bowel syndromes so as to preclude adequate oral nutrition. Although TPN carried out for several weeks may heal and temporarily close abdominal wall fistulas, patients relapse promptly on resuming feeding by mouth and the fistulas reopen.

Enteral Nutritional Therapy

I see no place for long-term management of ulcerative colitis with the use of nonresidue oral liquid-formula diets, such as those I have already mentioned. I am, on the other hand, well aware of our British colleagues' enthusiasm for the use of other forms of diet in the primary mode with or without any drug therapy. I simply have not found elemental diets either acceptable to my patients or their use effective. I must admit that my clinical trials with these approaches has been rather limited, so my opinion can be accused of being a parochial view. However, one can see how elemental diets could favorably influence bowel inflammmation by reducing the antigenic load profered the small intestine and colon.

17

The Role of Psychotherapy

Along with the continuing increasing interest and research in the etiology and therapy of IBD and the continuing tidal wave of papers on the quality of life with IBD and the methods for measuring it, I have been impressed by the disappearance of the emphasis over the last half-century on the psychological factors of these disorders. I cannot find a single mention of the subject in the index of the two leading texts devoted to IBD! Yet, when "psycho-somatic medicine" was the rage, ulcerative colitis was one of the big psychiatric disorders, along with hypertension, asthma, and peptic ulcers. This shift in emphasis is surprising at a time when the neural and biochemical interactions between brain and gut are a hot research area. Perhaps better drugs and surgery account for the difference. When I look at the problem from the point of view of etiologic factors predisposing or inciting IBD inflammatory mechanisms, there is little scientific evidence for the concept. In my own experience, a single traumatic event has rarely precipitated the onset of acute ulcerative colitis. I can recall but two. One was a patient mugged in

Lincoln Center Plaza. Another was caught in a retail shop during an armed robbery. I have the naive impression that among my young patients with Crohn's disease there are more children with divorced parents than in ulcerative colitis, but we live in a divorcing era and my impression may be quite erroneous and biased. Whatever gastroenterologists think they know grandmothers know that "nerves cause colitis."

Whatever the physician's instinct or bias is regarding the etiologic role of emotional factors in the life of the patient in the course of IBD, every physician needs to formulate a coherent stance regarding psychotherapy in the treatment of his or her patients. At the very least almost all patients, if not all relatives and friends, raise the issue. It is a current cliché that psychological and emotional factors play a role in the management of all illnesses, especially those of a chronic recurrent nature. But in IBD this problem seems to play a greater role if only because we have no real handle for understanding the basic nature of the disease and because of our intuition that there must be some relation between gut and brain. I do not know whether disturbances in the unconscious life of the patient lead to IBD, but I do believe that emotional turmoil and stress, both physiological and psychological, have a deleterious effect on the patient and his or her illness.

In order to prescribe individual psychotherapy, group and/or family therapy, and especially antidepressant medication, the physician must possess a sensitive sophistication and be able to diagnose and recognize psychological disorders in association with IBD, especially to recognize depression and be able to direct those patients in need, to sensitive and knowledgeable counselors who preferably have a special interest in IBD. I try especially to remember Thomas Almy's wise words written more than 27 years ago: "the life situation, the emotional state, the dependency needs and the psychologic defense mechanism of patients with IBD have come to constitute a necessary part of the *clinical description* of the illness" (Almy's emphasis). After this kind of clinical description of the individual patient, our therapeutic goals and measures cannot but be more widely ranging and compassionate.

In my observation, more psychological factors appear to be at work in the patient with ulcerative colitis than in the patient with Crohn's disease, and the key word here is "appear." On the surface, the patient with ulcerative colitis appears to be more sensitive to stress and the psychological defenses, especially ego strengths, seem weaker.

The dependent nature of the anxiety-ridden character of these patients with ulcerative colitis reveals itself in a characteristic behavioral pattern, which I observed over the years and which I call "one last question, doctor" syndrome. After the patient has left the consultant, or the physician has left the patient's bedside or the hospital room, the patient returns or follows the physician out to the door, or sends after for the physician to return to answer one last question. I have not conducted a statistical study, but this occurs almost overwhelmingly with ulcerative colitis patients in contrast to Crohn's disease patients. It is a reflection of their underlying anxiety.

On the other hand, some observers are not convinced about the question of more psychological factors operative in ulcerative colitis or indeed that there is more psychogenic factors at work in ulcerative colitis patients than in the control population, and they stress the part that depression plays in Crohn's disease. It is not difficult to understand that an illness like Crohn's disease occurring at critical times in the lives of young patients can be a heavy burden for them to bear. Further, since we do not know all the factors causing IBD, this does not mean that we may not be able to treat these disorders with this modality. Some treatments were discovered long before the causes of the diseases were found out. The question is not one to be settled theoretically. The real question is whether psychotherapy helps cure the disease. If it does not cure the disease and its attendant psychological stresses, does it help? I am not convinced that formal psychoanalysis carried out over long periods of time by a rigorous analyst has cured any of my patients. Much more helpful to me is to seek consultation with a well-trained psychotherapist interested in IBD, especially if the patient or the physician sense or suspect that the patient's reaction to the many problems of living with this disorder may play a part in the symptoms, in perpetuating the disease or triggering recurrences. In that case, psychotherapy and especially counseling, even on a superficial level, may help the patient to come to terms with this type of severe handicap. How to handle the realistic problems of work, education, girl or boyfriend, or spouse helps to improve the quality of life, but I caution the patient not to expect miraculous results and then fall into despondency and despair when a cure does not occur. Everyone, especially patients with these illnesses, needs to learn ways of handling tension in a constructive fashion. Tranquilizers, in my opinion, except in the short run do not work. More potent and complex psychotropic drugs, the mood elevators, require their prescription and use by skilled practitioners. Relaxation techniques, either in the form of exercise, yoga, TM, music, or psychotherapy, are all different approaches to the same problem. I have found that the patient must play the major role in making the decision to use them.

18

How Effective Are Our Current Drugs?

Does Meta-analysis Help?

I have in Chapter 14 outlined the "natural history" of untreated IBD insofar as it can be calculated from the placebo arm of controlled clinical trials of mild to moderate ulcerative colitis or Crohn's disease in an outpatient ambulatory setting. No such information exists regarding severe ulcerative colitis or Crohn's disease. In Chapter 13 on prognosis in IBD, I have outlined the information my colleagues and I have calculated of the likelihood of any patient with mild to moderate CUC or Crohn's disease going into remission or improving and the chances of maintaining that remission or improvement at present.

It remains for this chapter to try to estimate the effectiveness of current medications in IBD to answer the question posed by this chapter title. Since these estimates are based on controlled clinical trials of single drug therapy, they are an abstraction from the daily practice of gastroenterologists since one drug is succeeded by another if a prior one fails and often clinicians use two drugs simultaneously. However, this kind of study may help to set standards of reference that can determine the relative value of newer agents as they are proposed for the therapy of both Crohn's and ulcerative colitis and we certainly are into an era of newer candidate drugs: including in addition to the flood of 5-ASA preparations, nonabsorbable steroids, modifications of orally administered steroids, hydroxycloroquine, Omega 3 fatty acids, and calcium channel blockers.

The question of efficacy in this rather simple approach is the therapeutic advantage or gain for the patient when any drug is used. The therapeutic advantage of any drug is thus considered as the difference between drug and placebo response.

The Therapeutic Advantage of Current Drug Therapy in Ulcerative Colitis

In 11 trials involving 468 patients using various criteria for success, single drug therapy for the induction of remission in mild to moderate ulcerative colitis ranged from 37 to 43 percent (95 percent confidence interval = 22 percent to 52 percent). In five trials of 343 patients, the therapeutic response at six months was 21 percent (95 percent confidence interval = 4 to 38 percent), but increased to 46 percent (95 percent confidence interval = 25 to 67 percent) in 12-month trials because of the decline in placebo responders with time.

In studies of severe ulcerative colitis meeting Truelove's criteria, in five of 295 patients in defined trials, induction of remission was achieved at 53 percent (95 percent confidence interval = 49 to 79 percent). Of the 205 patients in trials where colectomy was a predetermined end point of failure, 35 percent lost their colons. Maintenance of remission was achieved in 54 percent followed for six months (95 percent confidence interval = 45 to 63 percent), but only in 35 percent followed for one year (95 percent confidence interval = 32 to 39 percent).

Therapeutic Advantages (TA) in Crohn's Disease

My colleagues and I looked at the placebo-controlled trials for inducing remission in Crohn's disease in 11 studies of 767 patients carried out for from six to 26 weeks. In achieving complete remission, single drug therapy yielded a therapeutic advantage of 10 to 29 percent over placebo (RD = 0.20; 95 percent confidence interval = 0.09 − 0.31). In achieving at least clinical improvement, the therapeutic advantage over placebo was 10 to 34 percent (RD = 0.23; 95 percent confidence interval = 0.08 − 0.37). When the drugs were analyzed separately, steroids had a TA of 29 percent, whereas azathioprine and sulfasalazine had a therapeutic advantage of 10 to 11 percent. In maintenance studies in the studies of 796 patients followed for 12 to 36 months, the therapeutic advantage persisted in 65 percent for one year, 28 percent for two years, and 20 percent after three years. No drugs had any advantage over placebo in maintenance of remission.

Summing Up

Our current drugs are effective in inducing remission and maintaining it in mild to moderate ulcerative colitis. They do confer a therapeutic advantage over purely supportive treatment. In severe ulcerative colitis, they effectively save lives and are less successful in saving colons. In Crohn's disease, our current drugs confer a significant advantage over supportive therapy in mild to moderate Crohn's disease with a clear advantage of steroids over other drugs when used alone. While most published studies reveal no therapeutic gain over watchful waiting in the maintenance phase of therapy, recent evidence favors the use of 5-ASA.

For maintenance of ulcerative colitis, I advise 5-ASA or its variants; for Crohn's disease, I have opted for 5-ASA maintenance therapy also.

To answer the question, yes, meta-analysis can help to sort out the bewildering data in the literature but it is of limited value in helping us choose the successive kinds or combinations of drugs in those patients who fail to respond to one.

The Future of Drug Therapy

I believe that all concerned with the medical therapy of IBD (Crohn's disease and ulcerative colitis) will agree that our current drugs leave much to be desired. They would also agree that these current drugs attempt to inhibit the biochemical effect of a multitude of mediators of inflammation. I am well aware that the abstractions of meta-analysis are far removed from the realities of current clinical gastroenterologic practice. Failure in the patient with either of these diseases to respond to Drug A leads to the exhibition of Drug B and failure of Drug B leads to the use of Drug C.

To this must be added the information that current drugs act at difference points of the inflammatory process, attempting to reduce the concentration of a number of mediators—leukotrienes, cytokines, prostaglandins, free oxygen radicals, tumor necrosis factor, and platelet activating factor—or blocking specific receptors of these mediators or the releasing factors or altering their formation by dietary change. This certainly suggests that many points along the inflammatory–immunological chain need to be suppressed if we are to improve the chances of therapy efficacy. Single drugs have not done the job.

Further, although few control trials have stratified their subjects according to initial attack and relapse, it has been clear since the classic study of Truelove and Witts on cortisone in ulcerative colitis that in this disease it was found that therapy was better in the first attack than in relapse episodes and this was found in severe as well as milder cases.

To this I also add that we should remember that the modern era in chemotherapy in neoplastic diseases began with the concurrent use of several drugs simultaneously or in sequence which attack the neoplastic process at several metabolic points. I believe that the immediate near future will begin with the premise that further advances and trials will depend on the use of several drugs simultaneously or in clear sequence acting on different loci in the inflammatory chain and this approach will be employed as early as possible in the first attack or episode. This kind of prediction is consistent with the emerging thinking that the synthesis of specific receptor antagonists or inhibitors of a single mediated inflammatory episode will probably not have important therapeutic consequences. Certainly the emergence of any entirely new class of drugs will receive a separate trial against the established drugs.

This prediction of the future use of multiple drugs simultaneously is consistent with other writers' ideas, as Daniel Rachmilewitz, for example, writes: "There is little chance that a specific receptor antagonist or an inhibitor of the synthesis of a single mediator will be of important therapeutic benefit."

VII

SURGICAL MANAGEMENT

19

The Decision for Surgery

There are points or nodes in the life histories of patients with IBD when a decision must be made regarding discontinuing medical therapy and resorting to surgery. Such decisions, which can occur at any point in the course of the illnesses with acute onset or unremitting persistence, are not easily made in diseases like ulcerative colitis where surgery in the past has carried with it, usually but not always, permanent ileostomy (although the use of the newer more cosmetic rectal-saving operations has sweetened the pill); in Crohn's disease, on the other hand, operation carries with it high risks of recurrence.

Decision for Surgery in Ulcerative Colitis

If one looks at all the patients with whatever extent of ulcerative colitis, one may ask how many will ever come to operation? As I pointed out in Chapter 13, reports from tertiary referral hospitals, such as my own, cannot answer this question. Only the better-designed and collected data of large catchment areas with total lifetime follow-up, as in the Scandinavian countries and Great Britain, are more useful in this context. Their records suggest that about 20 to 25 percent of patients with ulcerative colitis will come to operation within 10 years and up to 30 percent after 15 to 18 years. This data, as I have already indicated, fits in with my own general impression.

Emergency Decisions in Ulcerative Colitis

Some complications of ulcerative colitis are so straightforward that one need not agonize over the decision to remove the colon.

Free perforation can occasionally occur, usually in a patient being treated for an acute severe attack of colitis and almost invariably these days while on large doses of IV or oral steroids, which makes the diagnosis difficult even to suspect. This is why daily or twice daily abdominal percussion for the presence of liver dullness is so important. There can be no quarrel with the decision to remove the colon in this complication. However, I remember on one occasion that a difficult colonoscopy done by a master for long-standing ulcerative colitis with a mass of pseudopolyps was followed by the presence of air under the diaphragm with obvious clinical symptoms and signs. Conservative observation, intestinal intubation, and antibiotics tided this patient over the crisis, both physical and emotional, without the vehemently rejected colon removal.

Massive unresponsive hemorrhage in ulcerative colitis does not result from one single locus, but from diffuse ulceration and inflammation and requires colectomy of the entire organ. I have always thought in such situations the operation of choice must be pancolectomy, including the removal of the rectal segment, because, on a few occasions, failure to remove the rectum led to persistent bleeding or recurrence of bleeding and the necessity of further urgent surgery in the immediate postoperative periods. Some recent experiences and the testimony of experienced colleagues has led me to take a more conservative stance and now I no longer request my surgeons to always remove the rectum when intractable bleeding is the sole indication for operation, thus allowing for a latter ileoanal pull-through operation.

Other Complications in Ulcerative Colitis

Toxic dilatation of the colon in my experience is not an absolute indication for colectomy in patients during severe acute episodes of ulcerative colitis. While this does increase the morbidity and mortality of the episode, most observers, including myself, prefer to treat the patient vigorously with IV antibiotics and steroids or ACTH for at least five to 10 days with careful monitoring, including daily flat plates of the abdomen and small bowel intubation. While at least half of the medical patients who recover from an acute episode of toxic dilatation ultimately come to colectomy, I do not use this as an argument for foreclosing a trial of medical therapy.

The suppurative complications of ulcerative colitis present some of the most difficult decisions we need to make in the course of managing patients with ulcerative colitis. A *walled-off perforation,* usually in the sigmoid colon and often in the course of an acute episode, poses the question of surgical intervention very clearly and delicately. Careful watching is needed. The development of signs of spreading peritonitis is a good indication that the physician cannot procrastinate further with long-tube intubation, IV support, steroids, and antibiotics.

Less Urgent Decisions

Cancer of the colon in ulcerative colitis as in any other individual is an absolute, but not necessarily an emergency indication for total colectomy and is discussed in Chapter 26, where the question of dysplasia and isolated adenomatous polyps are also dealt with.

Prophylactic Colectomy

The question of surveillance is discussed in detail in Chapter 26, but up to the present I have not suggested colectomy prophylactically because the patient has had universal ulcerative colitis for more that nine to 10 years of left-sided disease for more than 15 to 20 years, simply because of these risk factors to prevent cancer in the future. I certainly use the clinical picture of the patient as a primary factor in making the decision, no doubt influenced by the length and extent of disease, even perhaps subliminally. Clear-cut verified and repeated biopsies with severe dysplasia, however, are in our current climate and will continue to be an absolute indication for colectomy. In this context, stricture of the colon deserves special attention. While benign strictures are much more likely to occur in Crohn's disease than in ulcerative colitis, benign strictures can and do occur in ulcerative colitis. More important, it is often quite difficult to decide that the stricture is really innocent. Symptoms due to mechanical obstruction make the decision for colectomy easier for us, but not for the patient.

I have been troubled by several recent instances of ulcerative colitis where colonoscopic observation and biopsies of the stricture and the mucosa on both sides have failed to detect the invasion of small cancers despite frequent follow-up surveillance endoscopy that has been performed at eight- to 12-month intervals. I intend to use CT scans and MRI more frequently in strictures in ulcerative colitis and endoscopic sonography if the lesions are within reach of the instrument.

Elective Colectomy

The most difficult decision we have to make in managing a patient with ulcerative colitis is the recognition that our trials of medical management have failed to induce significant remission in the course of the patient's illness or to improve the quality of his or her life. Frequently this involves the recognition of the undesirable side effects of our steroid therapy; cataract formation, gastroduodenal ulceration, osteoporosis and/or osteomalacia, and psychotic breaks are no longer tolerable prices we can ask our patients to pay even for sustained remission. All these indications may be grouped together as the patient with intractable colitis.

The extraintestinal manifestations of ulcerative colitis also influence the question of colectomy. The whole spectrum of extraintestinal manifestations of IBD are discussed in Chapter 28. Here I only want to discuss *pyoderma gangrenosum*, *arthritis*, and *sclerosing cholangitis*. Recurrent pyoderma gangrenosum can be a most serious complication of chronic ulcerative colitis, and the question arises: Is this an indication for surgical intervention? In most cases, the pyoderma is part of a severe form of ulcerative colitis, and in these instances failure to respond to our general and local forms of therapy certainly makes colectomy an acceptable therapy of last resort. My experience is that when the patient's underlying colitis is severe and active, healing begins promptly after the colon has been removed often during the same hospitalization as the surgery and proceeds quickly to resolution. When the colitic process is milder or more moderate, healing of the skin will ultimately take place, but may be much slower and can require up to six to 12 months for completeness, but almost all patients with pyoderma gangrenosum get better after the colon is removed. Unfortunately, there are a few instances, and I have seen one, when pyoderma recurred after total colectomy. I have seen pyoderma which continues even when almost all, but not completely all, of the rectal colonic mucosa has been removed in preparation for a second-stage pelvic ileal pouch with planned mucosal stripping. Removal of the residual inflamed mucosa resolved this problem.

Axial forms of arthritis or sclerosing cholangitis unfortunately do not parallel the degree of colitic activity and do not improve following colectomy. I have seen only one patient for whom I hesitantly suggested colectomy for severe recurrent serologically negative peripheral arthritis following failure to respond to steroids, 5-ASA in its various forms, and azathioprine, as well as a course of methotrexate.

Decisions for Surgery in Crohn's Disease

We all share the basic feeling that, in contrast to ulcerative colitis, patients with Crohn's disease present us with more difficult problems and more difficult decision-making situations and we expect that more patients with Crohn's disease will come to operation than those with ulcerative colitis. In some series of Crohn's disease of the small bowel, up to 65 to 85 percent of patients came to surgery in long-term follow-up studies. What about Crohn's disease of the colon? In Cleveland, about 58 percent of patients with Crohn's of the large bowel came to operation in the course of their long-term follow-up.

When are patients with Crohn's disease operated upon? Surgery is recommended for free perforation, walled-off perforation with abscess formation, obstruction, varieties of fistulas (intra-abdominal, abdominal wall, and perianal), hemorrhage, toxic dilation of the colon, and cancer, as well as medical failures, including the inability to be weaned from steroid therapy.

Emergency Decisions in Crohn's Disease

Free perforation of the bowel into the peritoneal cavity is fortunately rare in this disease, probably due to the tendency of Crohn's disease to extend into contiguous, agglutinated bowel, but when it does occur, it is a clear-cut indication for prompt surgical intervention. At times the perforation can occur in an area of normal small bowel just proximal to a narrowed stenotic area. Massive, or rather continuous, *intractable bleeding* is another but clearly much rarer indication for prompt operation.

Obstruction in Crohn's disease is discussed further in the section on obstructive problems and is usually not complete and often is associated within localized disease and/or abscess formation. However, obstruction may be solely mechanical in nature due to stenotic disease or often from an adhesive band in anyone who has had a prior laparotomy. On occasion I have observed episodes of incomplete obstruction converted to complete obstruction by tough foods becoming impacted in moderately severe stenotic areas—corn on the cob kernels, firm Chinese vegetables, a peach pit, popcorn, several colossal stuffed jumbo olives, and a big meal of raw vegetables, including broccoli. Most, but not all, of these episodes were managed conservatively with long-tube intubation for resolution of the emergency situation. A few required operative intervention.

Less Urgent Decisions

Although most *cancers of the ileum* in the setting of Crohn's disease are discovered accidentally at laparotomy, we are now beginning to make the diagnosis before surgery more frequently. A clinical flareup with the discovery of a mass after a long period of remission or long after a resection has helped me in making this diagnosis and alerting me to this possibility. Cancer of the colon, of course, is an absolute indication for surgical intervention, though not in an emergency setting. The appropriate cancer operation is in order without regard to the presence of Crohn's disease except as it effects the safety and extent of resection of the bowel.

Suppurative Complications in Crohn's Disease

Localized abscess which raises the question of surgical intervention is conventionally treated by parenteral antibiotics and in most cases, unless the abscess is quite small, does not respond to this treatment. In my experience, more is usually required than drainage of the abscess and the best results follow resection of the involved segment of the Crohn's diseased bowel. I have noticed recently an increasing tendency in the presence of a localized abscess or tender inflammatory mass associated with small bowel obstruction in this disease for clinicians to resort to IV steroids, hoping to suppress the inflammatory, as well as the mechanical aspects of the stenosis. I fear the use of steroids in this context as increasing the danger of extension of the suppurative process. In essence, what we might call infected intestinal obstruction rises from a localized inflammatory reaction in the wall of the gut and treatments of steroids and antibiotics resolve this temporary obstruction; but long-standing stenotic obstruction is due to recurrence and fibrosis; in these only a mechanical solution to a mechanical problem will do.

Abdominal wall cutaneous fistulas present another indication for operation. I have not seen cutaneous fistulas heal with our current medication, including use of immunosuppressants (azathioprine and/or 6-mercaptopurine), although others have presented evidence regarding their effectiveness. While closure of a cutaneous fistula may occur with complete bowel rest and total parenteral nutritional support, the fistulas invariably, in my experience, reopen as soon as feeding is resumed. Enteroenteric fistulas usually including the ileocolonic variety are not absolute indications for surgical intervention. Ileosigmoidal fistulas do not heal with current medications, but may be tolerated for long periods of time. Enterovesical fistulas, because of the erroneously presumed risk of ascending urinary tract infection, are usually considered an indication for prompt operation. However, fistulas to the urinary bladder can be tolerated for many years. Indeed, some colleagues have been managing this type of fistula by prolonged antibiotic therapy. Although I have had these fistulas operated on, even if the disease required ileostomy and colectomy, we might consider a more prolonged period of medical therapy. I have delayed surgical intervention in enterovesicular fistulas unless there is marked urinary difficulties from the passage of particulate matter per urethra.

Toxic dilation of the colon is seen and can occur in Crohn's disease, perhaps not as frequently as in ulcerative colitis. In some instances, as in ulcerative colitis, colectomy may be required if the patient does not respond to medical therapy.

Elective Surgery in Crohn's Disease

The ideal patient for surgical intervention in Crohn's disease is one with a mechanical obstruction of intestine and minimal evidence of inflammatory activity however measured or defined. In this situation, resecting the stenotic area or stricturoplasty if feasible is the treatment of choice. Recurrent obstruction with a suppurative focus is another clear-cut indication for surgical resection.

The *medically intractable patient* is, of course, the one whose Crohn's disease includes disabling symptoms that persist and who remains unresponsive to our current panoply of drugs—sulfathalazine, mesalamine, steroids, metronidazole, immunosuppressants (azathioprine or 6-mercaptopurine), as well as cyclosporine and TPN. Equally important is the patient whose disease is suppressed satisfactorily with steroids but only at a prohibitively high dose with its potential risks. Here we must operate despite the risk of recurrence of disease and its attendant symptoms. I may have the skewed view of a tertiary referral center long a pioneer in Crohn's disease, but it is my impression that the number of patients with medical failure in Crohn's disease has not apparently been reduced by the panoply of current drugs. Thus as many patients, it seems to me, come to surgery as ever.

There is one other rather curious indication for surgical intervention in localized Crohn's disease, which I have been involved in on at least two occasions. Two males with ileal and ileocecal Crohn's, one in his twenties and the other in his fifties, had repeated episodes of sepsis with organisms clearly arising in the gut which was the only portal of entry without local abscess formation. Both had had no further episodes following ileocolonic resection. This is a rare situation but worth bearing in mind.

20

The Choice of Operations

Operative Choices for Ulcerative Colitis

The problem is not the use of pancolectomy to "cure" ulcerative colitis; no other operation is available if the patient requires surgical intervention. The options turn on whether a permanent ileostomy will be required or whether it may be possible to restore intestinal continuity and preservation of the anal and rectal sphincters. I do not underestimate the importance of body image and the physical and psychological impact of an ileostomy at any age in both sexes, but I do stress the fact that restoration of health should be our aim and certainly is the patient's first aim of treatment. Cosmetics, important as they are, are secondary considerations.

In preliminary discussions with the male patient, we need to discuss candidly but reassuringly the fear or likelihood of impotence, stressing that it is probably much reduced when an operation involving the rectum is done electively, not urgently, and that these operations are not cancer operations done with wide dissection and finally that the risk is really very low. We ought to warn our patients that our surgeons in the present litigious climate tend to overstress this potential hazard to avoid law suits. I personally have had only one patient complain of impotence in more than 45 years of looking after patients with IBD coming to operation.

It is wise I believe in preparing the patient for the surgial method of choice to distinguish between operations performed in an urgent or emergency setting and those that can be carried out in a more elective atmosphere. For our patients who are sick, whose surgery is of an emergency or urgent nature, or who are chronically depleted and have been on steroids for a significant amount of time, ileostomy and subtotal colectomy leaving the rectosigmoid behind as a sigmoidal fistula or closed over as a Hartmann pouch is my unequivocal clinical choice. In these sick patients, especially young unmarried individuals, I do not think we should use the newer cosmetic operations as a bribe for the needed prompt operation. This kind of subtotal colectomy, leaving the rectosigmoid intact, allows us further time for realistic alternatives.

Choices for Ileostomy

For at least 50 years, the classical standard and most satisfactory operation for ulcerative colitis has been colectomy with proctocolectomy or subtotal colectomy combined with the Brooke ileostomy; and for most individuals this combination provided an extremely satisfactory quality of life, although with a price to be paid—life with an incontinent ileostomy.

The Continent Ileostomy of Kock

Since my earlier edition of this book, written in 1984, I am no longer in a holding position regarding this operation. I no longer advise it. Of my personally observed 50 patients with this operation, the twins with familial multiple polyposis have done splendidly. The remaining 48 individuals have had more than their share of problems. A few have found the necessity of frequent intubations with a catheter they carried about repugnant. One of them had the pouch removed within a year because of leakage. Twenty percent have been hospitalized one or more times for a revision of the pouch. None have had the troublesome condition of inflammatory changes in the pouch ("pouchitis" in our barbarous jargon), but I have seen several instances of other physicians' patients not always responding to medical therapy—sulfasalazine, local steroids, oral steroids, and metronidazole. One of my patients who had continent ileostomy because her Crohn's colitis was considered preoperatively to be of the ulcerative colitis variety requested removal of the pouch for systemic manifestation of pouchitis, including erythema nodosum; no granulomas were found in the resected ileal pouch. In one difficult case, problems have dragged on for several years. For these reasons, I am no longer recommending this operation. Certainly, the rectal-sparing operations, complex as they are, with their own complications, are more attractive at present for those individuals for whom the cosmetic and psychological problems of ileostomy are overriding.

With the increasing publicity and patient awareness of the development of rectal-sparing operations and continence, some younger patients who have had a pancolectomy and Brooke ileostomy in the past are anxious for a trial of the continent ileostomy of Kock. If they clearly have not had Crohn's disease and are otherwise well, this certainly is a reasonable alternative if approached by the patient with an open mind and open eyes, especially after first hand discussion with patients, some with good and some with poor outcomes of their continent Kock pouch ileostomy. Although I know that some surgeons have performed this operation with success in patients with Crohn's disease in the past after a long disease-free interval, I am reluctant to advise this in their cases, but some patients have gone abroad to have the operation in this context.

Operations for the Preservation of Intestinal Continuity and Rectal Defecation

I should begin this section on operations that preserve or restore intestinal continuity with one reservation. I saw little place for any operation except the Brooke ileostomy with a pancolectomy for individuals with high-grade dysplasia or carcinoma of the colon, but I have been weakening on this point.

The *ileorectal anastomosis*, since it was popularized by Stanley Aylett 30 years ago, has gone through successive waves of interest and disuse and has been revived again by Vincent Fazio most recently. It has obvious attractive qualities. It seems to be an easier operation than most. There is no stoma and no risk of impotence. Frequent loose stools and the need for antidiarrheals for at least the first year are acceptable to most patients, but the most complicating situation is the very real risk of carcinoma in the residual 15 centimeters of rectum left behind in this operation. It is this latter risk that has kept most clinical gastroenterologists from advocating this approach for their patients and I share their anxieties about the risk. There are however two situations where my patients have had an ileorectal anastomosis performed. One arises when the surgeon prepared to perform a pelvic pouch and pull-through operation finds this cannot be accomplished because of the shortness of the mesenteric vascular pedicle of the ileum. The ileorectal anastomosis seems a fair way of satisfying the patient's desire of preservation of intestinal continuity provided that the risk of leaving the rectal mucosa behind has been fully explained to the patient beforehand and the need for continued surveillance. The second situation is the discovery of liver metastases in a patient with colonic cancer who is scheduled to have an ileostomy and pancolectomy. An ileorectal anastomosis is probably a reasonable, humane alternative operation.

The Pelvic Pouch and Ileoanal Anastomosis

The muscosal stripping with pelvic pouch and ileoanal anastomosis represent the most imaginative advance in surgery of ulcerative colitis in my lifetime. I make this statement well aware of the fact that the surgeons are developing increasing skills, subtle nuances in technique, and that the quality of life with the pelvic pouch still presents problems. When it is successful, the ileoanal anastomosis with pull-through pelvic pouch offers intestinal continuity, preservation of the rectum, and avoidance of impotence in the male, and has no problem in the healing of the perineal wound, which may be slow with proctectomy. This must be balanced by the patient's acceptance of the need at least for the present of a two- or often three-stage operation, the surgical complications of pouch, and leakage at the cuff of the pouch, temporary, especially nocturnal, incontinence during the first six- to 12-month postoperative period, and the development of a stubborn complication of inflammation of the pouch, as well as the development of a stricture in the pouch. I am also aware that many of my distinguished colleagues feel that we have not yet seen the long-term complications of this surgical attempt to make the small bowel do the work of the colon. The ideal candidates are the well-motivated young persons whose surgery has a better chance of success if they are not depleted. Slowly, but with increasing experience, my patients' ages have gone beyond the late fifties and into the sixties. Low-dose steroids do not present a problem, but this operation in my opinion is not for patients at the time when they are on high-dose steroids or on other immunosuppressant medications and certainly not to be done as an emergency procedure. I believe all patients should be informed about the possibility of this operation, given its details, and urged to discuss its success not only with the surgeons who do the operation every day of the week, but equally important to obtain firsthand information from patients who have been through and lived with the operation. The Ileostomy Society, as well as the Crohn's and Colitis Foundation of America, I have found to be most helpful to my patients.

Yet, for the mature individual tired of his or her chronic illness, anxious to return to good health promptly, to resume career, education, and life, the one-stage Brooke ileostomy with pancolectomy is the most direct route to this goal.

To summarize, patients with ulcerative colitis whose quality of life cannot be maintained on current medications, especially if they cannot be controlled without steroid maintenance, are candidates for surgical intervention. Extraintestinal manifestations of the disease, including pyoderma gangrenosum, are rarely the sole indication for operation. Cancer of the colon obviously calls for surgery, as well as the presence of high-grade dysplasia in those individuals with more than eight to 10 years of universal disease. All candidates for pancolectomy should be given the opportunity to consider restoration of intestinal continuity and rectal preservation with the ileopelvic pouch and ileoanal anastomosis. For those who are desirous of returning promptly to health with a single operation, a Brooke ileostomy and pancolectomy is the operation of choice. Ileorectal anastomosis remains a much less desirable alternative except for palliation of cancer of the colon with liver metastases. For the few individuals whose operation will not allow for the promised ileal pouch, ileorectal anastomosis remains a fallback option, but with the need for lifelong cancer surveillance of the residual rectal mucosa.

Operative Choices for Crohn's Disease

In discussing the choice of operations open to patients with Crohn's disease who have failed to improve or go into remission with our current extensive panoply of medications and especially those who cannot be weaned from steroids with other immunosuppressant medications, it is prudent to warn the individual patient that surgery is but one part of his or her lifelong management problem. In this way, much disillusion will be prevented for the sufferer of this chronic and at present incurable illness with its eventual tendency to relapse and recur. Every patient operated on will have a recurrence if she or he lives long enough, but this recurrence may be only microscopic or radiographic and usually with time responds to medical treatment with significant improvement. Ideally one would like to have the patient operated on when the problem is mainly a mechanical one: obstructive stenotic bowel with mechanical symptoms, and minimal inflammatory activity. Equally important, the aim is to remove the absolute minimum of bowel to prevent small bowel insufficiency.

Operations for Small Bowel in Crohn's Disease

When those patients whose long history (usually nine to 14 years in my experience) has led to the development of one or more stenotic areas in the small bowel, the choices are few and clear. For disease involving the terminal ileum or the ileocecal area, local resection with direct anastomosis of adjacent grossly normal small bowel and colon is the uniformly desirable operation. In the presence of several or many stenotic areas in the small bowel, stricturoplasty as pioneered by Emanual Lee and John Alexander-Williams has been a significant advance in preserving small bowel mucosa and many stricturoplasties in the same individual can be done at one laparotomy with no greater risk of recurrence than with resection.

Most clinical observations have confirmed the fact that the resection does not need to be a wide one with lymph node resection as in cancer with no lesser recurrence rate than a limited resection. In jejunitis, which may lead to several stenotic areas with dilatation between them, multiple resections with skip areas are feasible and desirable. In dealing with ileojejunitis with multiple skip areas throughout the entire small bowel, the surgical attack must be limited and becomes a compromise between the need to resect all diseased tissue and the need to leave as much normal bowel as possible. In many instances, the patient needs to be explored and the most stenotic areas either resected or subject to stricturoplasty leaving some disease behind. I have seen some patients explored early in the course of the disease by experienced surgeons and found inoperable who, when explored 10 to 15 years later, now were found to have limited areas that could be removed and treated by stricturoplasty.

My own experience, as well as those of others, leads me to believe that the recurrence rate is higher where an anastomosis is performed. But, if intestinal continuity can be restored, I certainly favor anastomosis despite the risk, even if an ileoproctectomy or ileosigmoidectomy must be performed with anticipated loose stools or the patient require an ileostomy at a later date. In contrast to ulcerative colitis, anastomosis of the small bowel to the colon or rectum need not be done only in the presence of grossly and microscopic normal bowel. In Crohn's disease, anastomosis of grossly normal small bowel to mildly or moderately inflamed rectosigmoid can be done and need not lead to an operative leak, nor am I convinced that the risk of recurrence is greater in these situations.

Is there a role for bypass in that resection? In Crohn's disease I favor resection of all abnormal tissue rather than bypass for many reasons, not the least of which is the danger of cancer and the difficulty of diagnosing it in a bypassed loop. However, if exploration leads to an unexpected discovery of an abscess and a resection cannot be done safely because the ureters are caught and may be damaged in an inflammatory mass, then a two-stage bypass operation or later resection may be the better part of valor.

There is also a place for diverting ileostomy in Crohn's disease of the colon as the first-stage surgery in the very sick patient, especially one with rectovaginal or perianal fistulae or hydronephrotic obstructed kidneys and ureters. The first stage leads to a subsidence of toxicity and a marked temporary clinical improvement, but it is important not to temporize too long and postpone the needed second stage of colonic resection with or without anastomosis. If a pancolectomy is to be done for Crohn's disease of the colon and rectum, the physician ought to warn the patient when the perianal area has been the seat of fistulas in the past that healing of the perianal wound may be tedious.

Options for Crohn's Disease of the Colon

While cancer of the colon and probably soon the recognition of high-grade dysplasia in the colon in Crohn's disease and the development of mechanical strictures clearly call for prompt intervention, the most frequent indication is the continuous intractible course of colonic Crohn's disease. Most patients cannot and do not tolerate the colonic variant for as long as they do small bowel disease. If ileosigmoid or rectal anastomosis are acceptable for colonic Crohn's disease even with some involvement of the rectum, the question of doing an ileoanal anastomosis with a pelvic pouch or a proctectomy with a continent abdominal Kock ileostomy is frequently raised. For the most part, the rate of recurrent Crohn's in these operations is so high that these operations are not done except inadvertently. Although it is true that occasionally a patient, who is later discovered to have had Crohn's colitis rather than ulcerative colitis, has an acceptable result of operation, the chances of this occurring are slight so that I do not advocate this approach. If the whole colon, including the rectum, is involved with Crohn's disease, especially if there is evidence of a rectovaginal fistula and more importantly considerable perianal disease is present, proctectomy and Brooke ileostomy is the operation of choice. In these instances, the physician is right to forewarn the patient that when a perianal area has been the seat of fistula in the past, healing of the perianal wound may be slow and tedious.

Options for Suppurative Complications of Crohn's Disease

In the presence of localized, walled-off, intra-abdominal abscess, the basic question is whether to do an open surgical drainage or to perform a percutaneous drainage under sonographic or CT scan guidance. While percutaneous drainage of intra-abdominal abscess is in considerable vogue at present, I have not been impressed with its value in Crohn's disease where most abscesses are secondary to localized perforation. For those with small bowel disease of the ileum, most abscesses are secondary and proximal to a stenotic area. In these cases I prefer my surgeons to enter the abdomen extraperitoneally and resect the involved bowel and abscess en bloc. I believe in this way we prevent recurrence of perforation and achieve adequate drainage of the abscess and do not run the risk of a cutaneous fistula.

Perianal Disease

My professional career has seen a 180-degree turnabout in the treatment of perianal disease. Incision and drainage was the slogan of my youth. Now fistulectomy and sphincterotomy with drainage of the intersphincteric abscess (now considered the chief etiologic factor), the so-called Parks procedure, is the rule of the day. I see this as a rational advance to recommend to my patients, but I am influenced by the degree of disease activity upstream. Ileal or other small bowel disease presents no problem. However, I am still a bit concerned with the healing of the Parks procedure in the presence of colonic or more proximal rectal disease. In the presence of severe anal disease, most often with perianal disease and fistula formation, one occasionally may be forced to recommend proctectomy with a sigmoid colostomy. With more extensive colonic disease, ileostomy and pancolectomy may be indicated. In the presence of much perirectal inflammatory disease, including an active rectovaginal fistula, a prior ileostomy followed later by colectomy may be needed in order to allow the subsidence of local vaginal and rectal infection and edema.

21

The Results of Surgical Therapies

The assessment of the results of surgical intervention in these disorders obviously must be made clearly in relation to the primary aims of our therapy. In both Crohn's disease and ulcerative colitis, the way to measure success is clear if an operation has been done for a life-threatening massive bleed, a free perforation, spreading peritonitis from a leak or walled-off perforation or abscess, or complete mechanical obstruction unresponsive to conservative management. The patient's survival is what counts. Hence, these indications and cancer in either of these diseases pose no complex problem in determining outcome. It is when we turn to patients operated on electively, because of intractability, that is, our failure to manage the illness well enough with the acceptable risks of our drugs, that the problem of assessment of success becomes more difficult. In ulcerative colitis, the aim is to "cure" the disease. In Crohn's disease, the aim is to improve the quality of the patient's life.

Results in Ulcerative Colitis

The immediate response to a total or subtotal colectomy in these sick, depleted individuals suffering from both the disease and the toxic effects of current therapy carried on too long, especially steroids, is the remarkably dramatic, prompt improvement in their general condition, appetite, color, and mood despite the inevitable distress and pain of this major operation and the patient's reactive depression, with its awareness of a possible future ileostomy life.

The more pertinent question is whether or not we have improved their general health. In my experience, with colectomies done for medical failure in ulcerative colitis, the overwhelming majority of patients are genuinely pleased by their restitution to a normal, healthy life.

If we first consider the results of therapy in those who have had a Brooke ileostomy and colectomy, whether total or subtotal, there is no question that these patients' lives have more problems than matched controls who have never had colitis or even those with ulcerative colitis who have never been sick enough to have been considered candidates for operation. But I feel this is a foolish comparison. The pertinent question to ask individuals a year after recovery is whether their lives are better. The answer in my experience is a resounding yes. The major exception is the patient who has had a severe form of ulcerative colitis for only a short period of time. While he or she is glad to be alive, they are not grateful for the prospects of an ileostomy life and a pancolectomy. The patients' expectations will be more realistic if they discuss life with an ileostomy with those who have had that experience which can be easily arranged by the chapters of the local ileostomy societies or the Crohn's and Colitis Foundation of America, and by pointing out in advance of elective surgery that at least one-quarter of patients with an ileostomy will have an episode of intestinal obstruction, and that perhaps a quarter of these will require a laparotomy.

Stomal dysfunction in my lifetime has been diminished tremendously with the Brooke form of ileostomy and use of modern appliances and instruction by the many enterostomal therapists available in any good medical center. Careful dissection of the rectum in these elective patients has led to no significant impairment of sexual or reproductive function in the women and only one man suffered from impotence in a long series of several hundred patients. This risk must be discussed candidly and the patient's anxiety responded to in a straighforward account of the individual physician's experience and the results of surgical consultants and their own patients, not by quotations from the literature.

Results of Surgical Intervention for Cancer of the Colon in Patients with Ulcerative Colitis

Cancer of the colon in patients with ulcerative colitis is discussed at greater length in Chapter 26, but some points need to be emphasized in the present context. Although the cancers of these individuals are often multiple and appear often at an earlier age than in the common sporadic tumor, the response to surgical intervention (colectomy) is no different from the ordinary cancer of the colon and is dependent on the stage of invasion of the colonic wall and its extension beyond the wall to regional nodes. An earlier fear that these cases are more "deadly" than the common noncolitic variety has not been borne out.

Results of Surgery in Ulcerative Colitis with the Newer Operations

The assessment of the results of surgery entered a new and interesting phase with the development of the continent ileostomy of Kock, the pelvic pouch and anastomosis with or without mucosal stripping, and the revival of the ileorectal anastomosis. Continuing experience with these approaches is resulting in a continuing reassessment of outcomes as each gastroenterostomist reviews his or her own experience. As I pointed out in Chapter 20 on choosing an operation, I no longer advise the continent ileostomy if there is any possibility of a pelvic pouch operation, except in the rare occasion that a patient with a prior Brooke ileostomy and pancolectomy wishes to try the alternative to a continent ileostomy. In one instance, a female patient, having had several unsuccessful attempts to "rescue" her leaking pelvic pouch persuaded Dr. Kock to substitute his continent ileostomy.

As I have also already indicated, I believe almost every patient facing pancolectomy should have the opportunity to consider having the pelvic pouch pull-through operation. Not only my surgeons, but I too have extended the age group for which the operation can be contemplated. The most likely candidates for successful operation are the young, motivated adults of thin habitus, but the age of 50 plus is no longer an obstacle. In my own series now of more than 50 patients, the majority of individuals have been satisfied with the quality of their life after surgery, and the reports of the larger series from the Mayo Clinic, Cleveland Clinic, and our own Mount Sinai Medical Center confirm this.

With increasing experiences and subtlety of operation, moderately severe surgical complications, especially leaks from the cuff of the in-flow track, have significantly decreased. Some degree of incontinence during the first postoperative year, especially at night, is the price most patients now must pay for the advantages of this operation. Continence during the day is usually achieved with the aid of antidiarrheal medications. The major medical complication facing all individuals with the pelvic pull-through operation, as well as a diminishing number who have had a variant of the continent ileostomy, is the development of the inflammatory condition within the pouch now barbarously labeled pouchitis: the induction of an acute mucosal inflammatory process resembling the original colonic one in the reconstituted ileal reservoir. Clinically difficult to control abdominal and pelvic discomfort, increasing number of evacuations, and loose stools with bleeding in the presence of systemic symptoms accompany this complication. It is my estimate that perhaps 30 percent of patients will have a mild form of pouchitis; a few unfortunate ones (10 percent) may have a more severe form. The striking fact that this complication does not occur in patients operated on for familial colonic polyposis, certainly suggests to me and to others that pouchitis may represent a manmade model of inflammatory bowel disease: secondary to bacterial stasis in an intestinal mucosa that has the genetic susceptibility to inflammation with the same immunological defect as existed in the colonic mucosa.

The majority of instances of pouchitis respond to the use of oral antibiotics such as metronidazole, and in some the addition of local therapy in the form of steroid instillations or the use of local 5-ASA enemas may be needed. One trial with sodium butyrate did not succeed. Some may require weeks to months of persistence, and even systemic steroids in rare instances. The small but more severe and persistent varieties of pouchitis raise other questions. The one most often considered is whether the recurrent inflammation now in the small bowel following removal of the inflamed colon may not simply be due to one's having had a patient with Crohn's disease of the colon operated on because of prior erroneous diagnosis of ulcerative colitis. This scenario has certainly happened in a few instances usually with individuals with "indeterminate" colitis.

However, a review of our own experience in a series of the more severe forms has shown that this common assumption of mistaken Crohn's disease of the colon is not the explanation for the overwhelming majority of instances. In the clinical study of these unfortunate individuals, men preponderate over women and have had many more extensive extraintestinal manifestations of their disease preoperatively than the group with mild pouchitis. Several puzzling but intriguing facts should be noted. Early experience with recurrent inflammation in inflow tract or in the pouch led to the abandonment of the use of the operation by most surgeons in any patient who might have had Crohn's colitis. However, it is interesting that some individuals whose colons are clearly of the Crohn's variety have done well postoperatively with the pull-through operation. Prudence, however, suggests at present that the ileal pouch is not to be done in clear-cut instances of Crohn's disease. For someone like myself, who has insisted on the separation of ulcerative colitis from Crohn's disease, it is disturbing that in some instances in which pouchitis led to the removal of the pouch and where careful review of the resected colon confirmed the original diagnosis of ulcerative colitis the pouch has revealed clear-cut mucosal granuloma systems.

A current clinical therapeutic dilemma focuses on this situation. The first stage of total colectomy of a planned ileopelvic pouch anastomosis for what in the past was the case of a usual ulcerative colitis revealed mild mucosal disease with superficial granuloma systems and not transmural in distribution. Our gastrointestinal pathologist, Noam Harpaz, informed me that he and others have seen other instances of this phenomenon. Whether to proceed with the creation of a pouch in the next stage remains problematic.

Errors in the choice of operation may be important. Some surgeons are beginning to develop the needed skills to perform pouch operations in one stage in elective cases which does not allow for the careful study of the resected colon that we have depended on up to the present.

A significant complication, although not frequent, has been the development of *stenosis* at the ileoanal *anastomosis,* which has required several of my patients to resort to a catheterization of the pouch in order to empty it completely. Some have even needed repeated mechanical dilatation of the stricture.

Results of Ileorectal Anastomosis in Ulcerative Colitis

Because of the fear of development of cancer in the 15 or more centimeters of rectal mucosa remaining in situ after the operation of ileorectal anastomosis, coupled with the problem of frequent loose stools and the at times difficult to maintain continence, I have not advocated this approach as a routine procedure, except, as I have indicated, where the surgeon cannot for technical reasons find a long enough vascular pedicle for a pouch to be placed in the pelvis. I have reserved it as a palliative operation when cancer metastases are present. My personal experience with ileorectal anastomosis for the usual patient with ulcerative colitis is limited. We await reports from Victor Fazio of the Cleveland Clinic for more firsthand long-term information regarding ileorectal anastomosis and its risk of cancer recurrence.

In helping the patient to choose his or her elective operation for ulcerative colitis, I am guided by the needs of the patient to be restored to a busy, active, and productive life. Certainly, in the older patient, the standard Brooke ileostomy with pancolectomy in one stage remains the operation of choice for most; but, the younger patient or the highly motivated one wishing to avoid an ileostomy, and perhaps all should be given the option of considering restitution of intestinal continuity with the pelvic pouch, ileoanal anastomosis.

I am well aware that some of my most esteemed colleagues throughout this country have the instinctive feeling that we will some day have problems with the large number of pelvic pouches from the fact that we have tried to make the small bowel do the work of the colon; but I think, at present, it is a reasonable operation that every patient informed of the available facts should have the option of thinking about and choosing.

Results of Surgical Therapy in Crohn's Disease

Since our goals are more modest in the treatment of Crohn's disease, we need to distinguish carefully the meaning of the variety of recurrences to balance these recurrence rates against the effects of the operation on the quality of the life of our patient.

Forms of Recurrence

In patients with Crohn's disease who are operated on (and in most series of patients, including my own, about 65 to 70 percent of patients come to bowel resection some time in their life), they all eventually will have a recurrence if they live long enough. The use of actuarial analysis, the employment of life tables, which takes into account the rates of recurrence in terms of the number of patients still at risk following any operation, definitively clarified this point. But all recurrences are not of equal clinical significance. Reoperation is a clear-cut endpoint, but this should not be the only criteria. We must separate histologic or pathologic recurrence from radiological or endoscopic recurrence and both from clinically significant recurrence. I count radiographic or endoscopic evidence, plus clear-cut flare of disease activity as significant evidence of this disease's recurrence in this group. It is this group of patients who come to reoperation that concern us in this discussion.

In this context, current studies, using early postoperative endoscopy, indicate a very high proportion of patients operated on have endoscopic evidence of recurrence, perhaps up to 70 percent; yet, if recurrences are defined clinically, then about 20 percent have such recurrence by two years, 30 percent by three years, and 40 to 50 percent by four years. Defined in terms of reoperation, these reoperations occur at a much slower rate, perhaps up to 30 percent by five years and possibly 40 to 50 percent by 20 years.

Our experience leads me to believe that the following factors do not significantly constitute risk factors for recurrence: the age of the patient at the onset of the disease, sex, presence or absence of granulomas in the resected specimen. Others have shown that the extent of resection as measured in terms of freedom from histologic changes at either end of the excised bowel does not influence surgical recurrence. This factor seems reasonable in light of the diffuse nature of the disease. (Frozen sections at the time of operation do not help the surgeon, in my experience, to determine how widely to excise the bowel.)

Much has been made recently of the initial location of the resected bowel, stakes being higher in ileocolitis, next in ileitis, and lowest in Crohn's colitis. A multivariant analysis has helped us clarify this problem since these recurrence rates are higher in patients with anastomotic operations than those with ileostomy. Thus the lower rates of Crohn's colitis seem, in our experience, related rather to the type of operation than to the site of the original pathological process, not that an ileostomy for Crohn's offers any secure protection against recurrence. Recurrence of Crohn's disease in ileostomy is a most distressing complication.

Further, actuarial life table analysis reveals recurrence rates in a more realistic form. Postoperative recurrence rates appear to be exponential, so for the population remaining at risk, the chance of recurrence remains relatively stable at about 8 to 10 percent per year no matter how much time elapses after the first resection.

Most striking in our patients was the diminished rate of recurrence in those patients with the longest periods of preoperative disease, shown especially in those with 10 or more years of symptoms of the disease. This observation, of course, may only represent the inherent characteristicless aggressive biology of the disease, but does lend support to the older metaphor of "waiting for the disease to burn itself out."

The discussion so far should *not* be interpreted as meaning that I favor ileostomy over any anastomotic operation which can be safely done in one or more stages. I would certainly prefer as a patient to have an anastomotic operation and take my chances of recurrence, than to have a primary ileostomy and total colectomy.

As yet there is little to say about the results of surgery in Crohn's disease for a complicating cancer, since no individual observer, including the present writer, has had extensive experience with the few patients who have been operated on with the prior diagnosis of cancer.

Multiple Operations for Crohn's Disease

Patients and physicians are understandably worried about the risk of further reoperation in Crohn's disease with the inherit risk of creating the malabsorptive problems of the short bowel syndrome. A little less than half of the patients with their first resection for Crohn's disease will require a second operation, and less than a quarter of these will require a third. So perhaps one might hazard a guess that 10 percent of patients having had a first resection will require more than one additional operation throughout their illness.

22

Postoperative Problems

Early and Late

A problem I have seen little commented on is the persistence of symptoms following subtotal colectomy for severe ulcerative colitis. Subtotal colectomy for this condition is not a new operation, but, in recent years, we have all advocated pancolectomy as the operation of choice. Now with the emphasis on preserving the rectal musculature and rectal segment for subsequent pelvic pouch and ileoanal pull-through operations, I have become more aware of difficulties related to the remaining small segment of ulcerative colitis. With the subsequent weaning from steroids, the course of the majority of patients operated on urgently has generally been a good one. However, malaise may reappear, which usually is ascribed to steroid withdrawal. The emergence of low-grade fever, even in the absence of mucosal or rectal bloody discharge, then raises the question of an overt pelvic abscess; CT and labeled leukocyte scans follow in a futile search for the fever of unknown origin. Finally, endoscopy of the rectosigmoid segment reveals the residual inflammatory process still persisting, although not as intense as preoperatively. This symptom seems to be characteristic of ulcerative colitis in contrast to Crohn's disease in which fecal diversion leads to a temporary subsidence of inflammatory activity.

In cases where the systemic symptoms are of a mild to moderate nature, local therapy in the form of 5-ASA enemas has been helpful. Resort to steroids is disappointing and depressing to the patient, as well as chagrining to the physician. In a few, unfortunately, the recurrence of pyoderma gangrenosum in the postoperative period has led to a more prompt mucosal stripping of the rectal segment with the formation of a pelvic pouch or to a change in the surgeon's plan to the performance of an abdominoperineal resection. When these surgical options are not appropriate (obesity, etc.), the question has been raised regarding the possibility of a mucosal stripping at the original operation or, in a subsequent one, without a pull-through operation. Mucosal stripping with preservation of the rectal musculature without the pull-through operation has been done in a few patients, but the later use of the musculature in my own experience has been limited. Yet, in several instances, this has seemed to be a useful compromise to avoid proctectomy and preserve the hope of a pelvic pouch in the future.

The fashioning of a pelvic pouch ileoanal pull through at the time of the original operation is being done more frequently at present as surgeons' experience have increased and undoubtedly will prevent the recurrence of this sort of postoperative problem.

Early Postoperative Problems

When patients develop fever after an operation for inflammatory bowel disease, surgeons hope the cause is trivial and above the diaphragm, gastroenterologists look for sources below the diaphragm, but both agree that flat plates of the abdomen sonography and increasingly now the use of CT scans, as well as leukocyte scans, are the investigative way to go, with the routine obligatory urine and blood cultures, as well as chest x-rays.

In my experience, the recognition that a *liver abscess* (or abscesses) is the cause of the fever is among the last postoperative complication to be considered. The early alterations in liver function tests are usually ascribed to anesthesia, operation, and prior TPN therapy. Physical examination is usually not very helpful and the very early sonographic or scanning techniques may not help at this time. Only continued observation will lead to the correct diagnosis, sometimes never confirmed, after the fever responds to the use of antibiotics, a choice often made arbitrarily.

Immediate Postoperative Intestinal Obstruction

The distinction between postoperative ileus and mechanical obstruction is often very difficult to make. Following operations in which the colon is removed or resected, failure to completely reperitonealize the bare areas of the abdomen early on can lead to adhesive processes with consequent obstruction. Occasionally, when the peritoneal pelvic floor has not been completely restored, a loop of small bowel may be herniated through such a gap and may be difficult to demonstrate with plain films or even with contrast material, unless they are taken with the patient in the upright position. This form of intestinal pelvic floor herniation is most likely to occur in my limited experience, if at all, long after the surgical intervention.

Late Postoperative Surgical Complications

Intestinal Obstruction Following Ileostomy

All patients who have been assured of "cure" following total colectomy and ileostomy for ulcerative colitis should be warned that they run the risk of intestinal obstruction due to an adhesive band, which will probably occur in 25 percent of them. Of these, possibly 25 percent will require surgical intervention. From time to time, the patient with a well-healed, well-functioning ileostomy following ulcerative colitis with some minor kinking or adhesions may precipitate an episode of obstruction by an indiscrete meal of poorly cooked vegetables and Chinese snow pea pods, broccoli stems or orange peels or persimmons.

In a few, the obstruction may occur in the abdominal wall musculature, which compresses the ileostomy outflow tract. Much less frequent in modern surgery since the perfection of the Brooke ileostomy is a narrowing of the stoma of the ileostomy at the stomal opening on the abdominal wall skin surface. In the past, serositis and frequently inflammation scar the outflow, but the significant improvement in the creation of an ileostomy has practically eliminated this sort of difficulty.

The problem of intestinal obstruction is more complicated in the ileostomy created for Crohn's disease, complicated further by the assumption that any patient showing obstructive changes in the small bowel following colectomy for inflammatory bowel disease must have had Crohn's disease. Further, for these patients, in addition to adhesions, abdominal wall defects, and serositis at the stoma, the principal problem may well be recurrence of Crohn's disease in the ileostomy. Although steroids and forms of 5-ASA may help in these situations, the majority of stenotic obstruction of the ileostomy in Crohn's disease will require ileostomy revision and often its relocation to the left lower quadrant.

Late Surgical Problems

Ulcerative Colitis Surgery and the Fate of the Rectal Segment

The presence of a rectal segment following operation for ulcerative colitis arises in two major instances at present. The first is the almost routine performance of a subtotal colectomy with Brooke ileostomy leaving the rectal segment behind either with a cutaneous mucus fistula or converting the segment into the Hartmann pouch, closing it over, and dropping it back into the pelvis. This is likely to be the case in my experience in patients operated on either urgently or as an emergency, especially in individuals depleted by illness, malnutrition, and steroid toxicity. In addition to the rare situation of systemic symptoms already discussed in the section "Early Postoperative Problems," the major problem in this kind of operation for the individual in the rectal or rectosigmoid segment of the colon left behind is the need for prolonged cancer surveillance. As is usually the case at present, this surveillance usually begins after eight to 10 years of disease and will need to be continued throughout the entire period the defunctionalized segment remains in situ. The major difficulty is the almost inevitable stricture and closure of the lumen of the segment, which prevents complete endoscopic surveillance. Inability to visualize the entire lumen should lead to prompt proctectomy. It is interesting that in a recent experience in which the rectal segment absolutely quiescent, for 10 years, was followed after rectal resection by the development of a pelvic abscess and a collection of fluid, which followed along an old sinus tract.

The second situation that leads to a residual rectal mucosa is the performance of ileoanal or ileorectal anastomosis in ulcerative colitis, an operation periodically revived because of the surgeon's inability to have a long enough ileum to reach the anal or rectal segment which was to be performed as part of a mucosal stripping, pelvic pouch, ileoanal pull-through operation. In both of these situations, the rectal mucosa must be closely observed, for dysplasia or carcinoma. The presence of these of course should lead to a prompt proctectomy. The diagnosis of cancer in a significant proportion of these patients has certainly continued to lessen our enthusiasm for the ileoanal and ileorectal anastomosis in ulcerative colitis without mucosal stripping.

In classic ulcerative colitis, the rectal wound usually heals well and promptly, especially now that surgeons are performing primary closure more frequently.

Problems After Surgery for Crohn's Disease

Following ileocecal resection for localized *ileitis,* development of a fistula almost always means that residual Crohn's disease has been left behind. Formerly, more common was the development of fistula following laparotomy and appendectomy for undiagnosed ileitis, which arose almost invariably from the inflamed ileum and not from the appendiceal stump. These fistulae have required a surgical attack of the diseased bowel and have not responded to current drugs.

In patients with severe perineal and perirectal disease who have either had a diverting ileostomy or colostomy and partial colectomy, the perineal disease usually but not invariably quiets down and does become asymptomatic. In some, the persistent Crohn's disease of the anus and rectum may flare up from time to time. Antibiotics, such as metronidazole, are quite useful in this regard as well as the increasing use of immunosuppressive drugs of the azathioprine or 6-mercaptopurine variety.

More disturbing is the failure of the patient to heal following total colectomy and abdominoperineal resection of *colitic Crohn's disease*. Fear of this complication leads us to delay resection of the rectal segment in patients with much rectal disease or a history of multiple perianal fistulas. Patients are also concerned about the development of impotence following operations in this area. While some such persistent sinus tracts following excision of the rectum clear up after debridement, local hygiene, and cleaning, some persist. Biopsy of the tract may also reveal histologic evidence of persistent Crohn's disease and often well-developed granuloma systems. Trials in a few situations in my experience with steroids and azathioprine have not been rewarding. Here the graft or gracilis flap operation is the answer and is successful, even in the presence of such histologic evidence of Crohn's disease. Experience with the local application of 5-amino salicylates is too recent to allow for any conclusion.

Persistent diarrhea, which occasionally follows a localized ileocecal resection with loss of the ileocecal valve, usually resolves spontaneously. But in some patients it persists and is annoying even after a reasonable wating period postoperatively. Here binding of the intestinal bile by cholestyramine or even the constipating antacids containing aluminum hydroxide are valuable and some patients may require their use for years.

Significant clinical recurrences of course often do occur. Usually, but not always, they are milder than the preoperative disease and the majority of the instances occur on the intestinal sides of the ileocolonic anastomosis in the "neoterminal" ileum. We really do not know how often the recurrence in the colon occurs in previously x-rayed or colonoscopic normal bowel, but its appearance is disconcerting. A more significant characteristic is the development of rectal Crohn's disease following ileosigmoidal or ileorectal anastomosis for Crohn's ileocolitis. Local steroids or even systemic steroids help, but frequently the anastomosis needs to be taken down to give the patient adequate relief. I have already commented on the performance of an ileostomy, especially in Crohn's colitis, may not protect against the recurrence of the disease in the distal ileum or the appearance of a peristomal fistula. This type of recurrence, even though it does not make the patient terribly sick, is often rather difficult to improve by current medical methods. Revision resection of the involved ileum and the often needed relocation of the ileostomy to the left lower quadrant may set the stage for the too often and too soon recurrence.

The Fate of End Ileostomies

When complications arise in patients with end ileostomies, they are almost invariably in those on whom total colectomy was performed for Crohn's disease. While it is true that some ileostomies performed for ulcerative colitis may develop inflammation at the mucocutaneous junction with resulting stenosis requiring surgical correction, the incidence of this complication is dramatically reduced by the universal introduction of the Brooke-type ileostomy. The commonest complication is the development of ileitis of the original Crohn's variety and the patient has been operated on for ileitis or for ileocolitis; but recurrence of increasing but as yet undocumented frequency can occur in the follow-up of pancolectomy for Crohn's colitis alone. Abdominal pain, an increasing volume of ileostomy affluent, ileal bleeding, systemic symptoms, anorexia and weight loss, as well as fever, develop insidiously often after years of wellbeing. Retrograde ileoscopy easily makes the diagnosis. In my experience the usual drugs, such as steroids and 5-ASA variants, used in Crohn's disease seem to help very little. Published experience on

Imuran® or 6-MP, methotrexate, or cyclosporine is lacking. The development of obstructive symptoms with concomitant development of a parastoma fistula or localized abscess is common. Attempts to treat the narrowing of the neoterminal ileitis of the ileostomy by mechanical dilatation in my experience is futile and often leads to local sinus tracts and peristomal abscess. Operative intervention with resection of the ileitis, along with drainage of the abscess usually requiring moving the stoma to the left lower quadrant, will solve the problem. It remains to be seen whether the newer 5-ASA compounds given routinely postoperatively for maintenance therapy will reduce this complication which fortunately is rare.

Intractable diarrhea upon the establishment of an end ileostomy is fortunately rather rare and, in my experience, is associated with the degree of the shortening of the bowel. In modern times one rarely sees individuals who require persistent TPN for maintenance nutritional treatment. In interesting and treatable mild variants of this syndrome of intermittent episodes of diarrhea years after the operation in Crohn's, patients almost always are uniformly suspected of having an inflammatory recurrence of the disease, but none can often be demonstrated by x-ray or endoscopy. However, in some, the presence of an incomplete or partial obstruction, kinking of the bowel often without pain and no clear obstruction on films of the abdomen, is due to external adhesions. Here ileoscopy through the normal ileostomy stoma revealed a sharp fixed angulation at 40 centimeters in one patient which responded nicely to lysis of the external band. It may be possible to demonstrate such a functional obstruction by the use of a radio-opaque marker study.

The Disconnected Bowel: Diversionary Colitis

The development of inflammatory changes in the colon or rectum following proximal diverting operations of a colostomy or ileostomy has been clearly demonstrated to my satisfaction, so-called diversionary colitis. This is difficult to treat with our standard therapy and the use of short chain fatty acid enemas introduced by Konrad Soergel has helped temporarily in a few of my patients.

Several with "reactive" arthritis secondary to the diversionary colitis responded promptly to closure of the ileostomy or colostomy and restitution of intestinal continuity.

The Phantom Rectum

We are all familiar with the patient who has had a subtotal colectomy leaving the rectum in place either as a mucous fistula or closed over and dropped back into the pelvis in the form of a Hartmann pouch. These individuals are occasionally startled by the sensation and need to defecate and the passage of a small amount of stool, mainly bacteria and desquamated mucosa from the disconnected rectum.

More difficult to treat are those few individuals who have persistent sensations following total pancolectomy or a secondary abdominoperineal resection. In some patients, pain in the perineum may herald the beginning of a small foreign body abscess, which slowly finds its way to the skin surface and is relieved by the discharge of a nonabsorbable suture.

In a few cases, when a total colectomy has been done for Crohn's disease, a fistulous tract containing granulomas may need to be dissected out. This appears to be the result of a residual bit of Crohn's disease. (The more common form of perineal persistent disease is the unhealed perineum of patients who have had much previous perianal disease prior to the total abdominoperineal operation.)

The more difficult, fortunately rarer, group are patients with well-healed perineal wounds who have the recurrent sensation of the urge to defecate and the sensations of a persistent rectum. These unfortunate individuals really do have the sensation of a "phantom rectum," equivalent to the causalgia of the patients with amputated limbs who experience the discomfort of their missing legs. After vaginal sonogram study and pelvic CT scans have ruled out residual disease in this area, this complaint remains a difficult problem to manage. I have resorted to transcutaneous sensory stimulating techniques and even to drugs such as Tegretol® with little success to report.

Pouchitis

The widespread recent use of a pelvic intestinal pouch in connection with rectal muscle sparing, mucosal stripping, and ileoanal anastomosis for ulcerative colitis has resulted in the appearance of many patients with significant and symptomatic inflammation of the pouch. Estimates vary, but I would guess that perhaps 30 percent of my patients have some form of this complication. While one may be tempted at first to conclude that this is a result of the presence of a form of diversionary colitis, a few trials with short-chain fatty acids in my hands did not help. The majority fall into two types in my experience: a mildly, less troublesome type and a second more persistent type much more resistant to treatment. Soon after the introduction of the pelvic pouch, "pouchitis" as it was then known was first ascribed as failure to diagnose the original colitic inflammation correctly: the operation had been performed on a patient with unrecognized Crohn's disease. More recent analysis, including our own at Mount Sinai Hospital, has shown that this is *not* the result of preoperative errors of diagnosis. Rather, it has been suggested that this may be a manmade model of the original colitic disease now in the small bowel occurring in the genetically susceptible individual. The much more transient less severe form of pouchitis does seem to be associated with bacterial stagnation in the pouch and usually responds to better emptying and the concomitant use of antibiotics, especially metronidazole. Some of my patients responded as well to local 5-ASA enemas, and few unfortunately required local steroid enemas.

The severe form of pouchitis, the more resistant pouchitis, presents greater clinical problems often with systemic symptoms. It is interesting that the males seem to be predominant and the patients are those whose preoperative courses were marked with more extensive forms of extraintestinal manifestations. When this form persists despite antibiotics and local 5-ASA therapy, I have been forced to resort to oral steroids. I intend to use the immunosuppressive drugs in the future to avoid removing the pouch as is necessary in a very few individuals. In one of these, resection of the pouch revealed a curious superficial mucosal granulomatous involvement not present in the resected colon. It is to be hoped that studies of this form of resistant proctitis with HLA typing and the presence or absence of ANCA determinations may help to segregate a specific subset of patients susceptible to this complication. Equally disturbing would be the presence of dysplastic changes occurring in the pouch, but few have as yet been reported.

23

The Quality of Life and Inflammatory Bowel Disease

The quality of life with IBD, the trendy Q word, is the current focus of a great deal of clinical study, writing, and interest. Judging from the literature, one might think that this area is a recent discovery. Indeed, in the words of one group, "quality of life with inflammatory bowel disease has not been well-examined" on the assumption that past studies based on enumeration of patient symptoms, together with the evaluation of their multiple laboratory and radiographic studies and alterations in the grading of endoscopic observation failed to evaluate properly important aspects of the impact of these disorders on the patient's lives. On the contrary, experienced and sensitive clinicians and even the most naive of practitioners have always used criteria of well-being or illness in their patients in judging the severity of their illness and in evaluating the success of their current treatments of them. What is of more recent origin is the attempt to quantify in some numerical fashion the end results of these judgments regarding severity of disease and any improvement in the patient's life and especially the recognition that the somewhat nebulous concept of "degree of well-being" or even "the quality of life" can be measured and quantitated.

When I suggested more than a decade ago that we could be more rigorous than simply impressionistic in attempting to answer the question of the quality of life after surgery for Crohn's disease, my own colleagues and collaborators were somewhat hesitant and skeptical about the possibility of doing this. But this decade has seen a considerable advance in interviewing techniques and technology in an effort to reduce observer bias. Much has and can be learned from the experience of opinion-poll takers who have reduced these errors in their field considerably.

But beyond improvement in the technology of measuring functional, emotional, and social criteria is the importance of the groups being studied and the precision of the questions we are attempting to answer.

We all would agree that it is better not to "have" ulcerative colitis or Crohn's disease even if we are "well" and do not seek medical care or treatment. Yet recently much effort has been expended on a questionnaire study of a validated type directed at such a population, even if we leave aside the question of the appropriate controls of such a study. We want and our patients especially want to know whether they will be "better" after any form of treatment, including expectant forms, than they are at present.

This necessitates that the population at risk be clearly identified, not merely as IBD, but as a specific illness, the starting degree of impairment, the type of medical, nutritional, surgical, and psychosocial support therapy that the patient is receiving if we want to move beyond impressions. Many of the recent studies often cited in this context do leave much to be desired regarding these specific points.

It is reasonable to measure the acute effects of treatment of the patients ill with IBD on their intestinal function—laboratory studies of inflammation and the imaging techniques used after short periods (two to eight weeks) in ulcerative colitis and longer periods (three to four months) for Crohn's disease. But for a true estimate of the impact of these illnesses on the lives of individuals with these unusual and chronic illnesses, the assessment must encompass long periods of time, especially if we are attempting to measure the efficacy of our long-term maintenance programs.

To be repetitious, we still lack the needed information we and the patients are interested in. Taken as a group, all patients with IBD, measured at a time when they characterize themselves as stable, with or without maintenance medication, probably feel they have some varying degree of functional impairment in their lives. But it is obvious that until this is more accurately measured, this global estimate is a guess. Equally obvious the degree of impairment of patients with a localized proctitus of ulcerative colitis is quite different from that of a patient with Crohn's colitis in a similar state of stability with immunosuppressant therapy.

In considering ulcerative colitis and its lifelong probability of recurrence, we need to know not merely the impact of the illness on the patient's social, educational, occupational, and psychosomatic activities, but equally important still to be established is the impact of the need for lifelong surveillance for possible cancer development and the need for long-term endoscopic and biopsy studies for dysplasia after the eighth or ninth year of illness. If the cancer fear and the risk of colonoscopy, as well as its limitations, have been eliminated by ileostomy and total colectomy, the lifelong history of an ileostomy still needs careful study not only from the point of risk of complications, including intestinal obstruction, but its psychosocial aspects. There are only a few that have not been the subject of casual impressions.

In regard to the quality of information on the quality of life with Crohn's disease, we have even less information on this long-term illness with its great incidence and impact on the young population at risk and with a high rate of needed surgical intervention.

In our study, our aim was to evaluate the quality of life after surgery for Crohn's disease using carefully defined questions regarding overall satisfaction, physical symptoms, social relationships, schooling and employment, recreational activity, sexuality, and body image in a group of 52 patients who had their first operation for Crohn's disease at Mount Sinai Hospital and could be followed up five to 10 years postoperation. Fifty-one were located and collaborated in the study which did, on the average, cover eight years postoperation. They comprised equal groups of Crohn's ileitis and Crohn's colitis with or without ileal involvement. They were queried about their six-month preop status, one-year postop, and interval situation. All had had enough preoperative dysfunction to warrant surgical intervention since 20 with ileitis and 24 with ileocolitis reported significant improvement. At the time of interview, only 15 percent of the ileitis patients and only 12 percent of the ileocolitis patients considered themselves as having some impairment or dysfunction. Despite the fact that 22 had extension of the disease, 20 had reoperations, and 14 required an ileostomy, 92 percent of the patients were satisfied with the surgical outcome and would have chosen surgery again. Only four (8 percent) of the ileostomy patients were dissatisfied, although they had significant improvement in the quality of their lives.

My current feeling is that we need even longer follow-ups to give our patients a more realistic view of their future. To this end we are now engaged in studying patients who have been living with Crohn's disease for 20 to 30 years and who have lived through the steroid, immunosuppressant era, as well as having been operated on.

Yet operations do contribute to the quality of our Crohn's disease patients' lives. Fear of recurrence so well developed among physicians and thus among patients should not obscure the need to improve patients' lives even if, in contrast with ulcerative colitis, we cannot as yet "cure" Crohn's disease. Less than 5 percent of my patients who have been operated on electively because of our inability to manage their disorders medically are disappointed with the outcome; (patients who have clinical recurrence soon after operation for Crohn's do not share our enthusiasm). I think that some of the distress with surgery is the postoperative bile-salt catharsis which is not always treated as vigorously as it needs to be with bile-salt binders. I do not think a recurrence that might require reoperation should obscure the fact patients have many times reminded me that "surgery improved my life for many years between operations."

VIII

SPECIAL PROBLEMS

24

The Problem of Fistulas

Fistula formation is a part of the natural history of Crohn's disease and at times a sequela of surgical intervention. An isolated rectovaginal fistula may, of course, occur on occasion in ulcerative colitis. More diagnostically puzzling is the occurrence of a sigmoidal urinary bladder fistula from a segment of sigmoidal inflammation which is the seat of associated diverticula, especially since sigmoid diverticulitis with bladder fistulization has been noted to be associated on rare occasions with some of the extraintestinal manifestations of inflammatory bowel disease, those involving skin and joints.

Intestinal Forms of Fistulas

If we consider the spontaneous fistulas that develop in the course of the natural history of Crohn's disease, the commonest in my experience is the enteroenteric, usually ileoileal or ileocecal with ileosigmoidal a close second. Fistulas of the jejunum and transverse colon are rarer. These are usually discovered during the performance of a small bowel study, while barium enemas rarely disclose them. A sudden increase in abdominal pain or the sudden development of sharp localized pain against the background of a recent quiet bowel, or a sudden short spurt of fresh rectal bleeding in a patient without a history of antecedent bleeding may alert us clinically to this possibility. The sudden onset of profuse diarrhea in a patient with known Crohn's disease and a background of increasing constipation also raised the possibility of the development of an ileosigmoid fistula.

With the passage of time I have been more and more impressed that the mechanical factor of bowel stenosis plays an important part in the pathogenesis of these fistulas along with the inflammatory component.

The presence of an ileoenteric communication does not in itself raise the question of surgical intervention. The ileosigmoid variety, which appears to be an attempt of the obstructed bowel to establish an intestinal bypass, always raises the specter of the extension of the disease into the colon. It is interesting and probably important from an etiologic point of view that at operation the surgeon usually needs only to resect or sew over the defect in the colon and only rarely does a localized sigmoid resection. While the ileocecal fistula does not represent in my opinion a clear-cut indication for surgical attack, the ileosigmoid does raise the question of operation a little more frequently. In the past I cannot recall any of these enteric sigmoid fistulas healing under a medical program. Now, advocates of the immunosuppressant 6-mercaptopurine believe that this agent can be useful in these instances. With the increased use of azathioprine and 6-MP and acceptance of lower toxicity, we are fairly certain soon to have acceptable evidence regarding their effectiveness in this specific context.

The enteric fistulas that burrow into adjacent organs are curious and interesting, the ileal-bladder one being the most common. When these are large and well established, diagnosis is no problem. Early on they are more difficult to recognize. In addition we need to ask specific questions of the patient with any urinary discomfort to inquire whether the patient is passing bubbles when urinating or has a sense of air coming through the penis or vaginal urethra. Some patients need to be instructed to urinate while in the bathtub to detect the passage of gas. Barium by mouth may reveal the large communications, but not always. Intravenous pyelograms and cystograms are also not very revelatory showing only minor changes in the bladder contour. Cystoscopy rarely reveals more than an edematous bleb on the dome of the bladder and may help the surgeon when he operates if he knows that the fistula is near the base of the bladder or trigone.

It is surprising that even gross bladder fistulas are well tolerated as I have observed for many years, even as long as four or five years without impairment of renal functions or ascending infection in the urinary tract. I know of no successful medical therapy for ileovesicular fistulas, but I believe some colleagues are attempting to treat them with long maintenance courses of antibiotics as did one patient of mine who is a physician. As with the sigmoid, most often not much needs to be done with the bladder except to sew over the hole. The need for resection in fistulas of the bladder is most rare. This, however, should not be considered, as suggesting that leaving ileal-bladder fistulas alone is the method of choice. I think they should be taken down. Indeed, most patients want them corrected because of the painful urinary symptoms.

Fistulas of the right ureter can occur, although I have never seen one, and the fistula into the right hip that I have seen was surprisingly easily handled by ileal resection and cleaning out of the dissecting tract. Fistulas of the right ovary and tubes in ileal disease are not so rare as to be reportable. Fistulas of the vagina occur almost invariably in women who have had a hysterectomy. It seems as if the removal of the uterus facilitates the course of the dissecting tract. An extremely rare situation is the extension of Crohn's disease into the pancreas by direct contiguity. A large pancreatic mass was appreciated in one such incidence with an abnormal duodenal swelling on the upper gastrointestinal series, this before the days of CT scanning and sonography. The pancreatic biopsy at laparotomy showed a form of chronic pancreatitis with epithelial granuloma formations. However, granulomas it should be remembered can occur in pancreatitis not associated with Crohn's disease.

There is an accumulation in the literature and in our own limited experience of the association of pancreatic involvement and Crohn's disease without this kind of direct extension of fistulization. In our patient with pancreatitis of unknown origin, the process was halted with a resection of the ileal component of Crohn's disease.

Colonic Fistulas

We now go on to colonic fistulization. The rectovaginal fistula of Crohn's disease or even the occasional rectovaginal fistula in ulcerative colitis may on occasion be difficult to diagnose. Large communications are easily seen on barium enema. Small ones suspected by the history of passage of gas from the vagina, malodorous vaginal discharge, or the intermittent passage of small specks of stool from the vagina can be quite elusive. Charcoal pills by mouth and their passage through the vagina is a simple, crude way of proving their existence. Even experienced gynecologists can have difficulty at times demonstrating them. Installation of a dye by enema into the rectum and the staining of a tampon placed in the vagina may at times prove the point.

Small communications can be born with equanimity by many patients for many years, but the large ones present pressing considerations for help. Medical therapy can reduce the discomfort and improve the esthetic problem, but rarely is curative. In ulcerative colitis, surgical repair of the rectal (anal) vaginal fistulas is hazardous, except in the presence of really quiescent disease and long remission and even here is not always successful. In Crohn's disease, rectal involvement is most often associated with multiple perianal and perirectal fistulas (discussed below) that makes surgical correction more difficult. Here the drugs used for perianal fistulization seem to be helpful: sulfasalazine, metronidazole, and immunosupressants of the 6-mercaptopurine and azathioprine group. But for the intolerable situation, protectomy seems to afford the only long-term solution.

Crohn's disease of the colon can have two characteristic abdominal fistulas—in the first, right-sided disease of the hepatic flexure burrows into the duodenum and, in the second, splenic flexure disease communicates with the stomach. Foul eructation and intractable diarrhea with attendant weight loss are the classic manifestations of these upper intestinal colonic fistula. Their demonstration depends, of course, on the size of the opening and often barium by mouth or by rectum is needed to reveal them. While ileosigmoid fistulas are most often easily recognized by colonoscopic, colonic fistulas to the duodenum and the stomach are rarely ever seen on endoscopic examination of the patient with Crohn's colitis. I have never attempted to manage these fistulas by any prolonged medical therapy, but intestinal antibiotics preoperatively improved the diarrhea and foul belching. It is my impression, at present, that this type of fistula is seen much less frequently possibly because Crohn's disease of the colon is tolerated by the patients for a shorter period of time than Crohn's disease of the small bowel.

Abdominal Fistulas

Two main sites of cutaneous fistula in Crohn's disease—the abdominal wall and the perineum—need to be discussed separately. Abdominal wall fistulas tend to follow preexisting tracks and are rarely seen in the life of unoperated Crohn's disease at present. There are however exceptions to this general rule. In the patient who has had an appendectomy years before without antecedent ileitis, development of ileitis may lead to a fistula in an existing appendectomy scar. The second exception is the development of a fistula through the umbilicus, as we have postulated through the obliterated ligament of terres.

Spontaneous perforation of an abscess to the abdominal wall from Crohn's disease is rarely allowed to take place today in patients with known Crohn's disease, but a small, walled-off abscess of ileum and/or colon may rarely and suddenly penetrate the abdominal wall. Most of these fistulas and areas of distress occur through abdominal wall scars in the sites of previous laparotomies or drain sites. The common and best known is the fistula that erupts at the site of the recently performed appendectomy in a patient just discovered at laparotomy to have Crohn's ileitis. This fistula rarely arises in the appendiceal stump, but from the ileitis itself. I believe that if a laparotomy is done for suspecting acute appendicitis and Crohn's of the ileum or other areas is discovered, a careful appendectomy should be performed avoiding trauma of any sort to the ileum, to prevent the problems of suspected appendicitis in the future in a patient with known Crohn's disease and right lower quadrant pain.

An interesting and most unusual variant of cutaneous enteric fistula was seen recently with the spontaneous rupture of an abscess postoperatively through the skin incision of a cesarean section in a young woman discovered to have Crohn's disease of the ileum at the laparotomy.

Cutaneous fistulas in the colon with Crohn's disease occur much less frequently, but they do occur most likely, in my opinion, arising soon after a colonic or ileocolonic resection due to either the breakdown of the anastomotic site or to a local abscess at the site of residual Crohn's disease that had not been completely excised at operation. While anterior abdominal wall fistulas may be reduced in size, complete and spontaneous healing is probably unlikely followed by any current medication, including immunosuppressant drugs. Long periods of intestinal rest with total parenteral nutrition often closes cutaneous fistulas. But these open promptly when oral feedings are instituted; they almost always require surgical attack.

Another interesting cutaneous fistula is one seen next to an end ileostomy done for Crohn's disease of the colon or ileocolitis. Almost invariably, they are not the results of mechanical defective ileostomy appliances, but of recurring Crohn's disease in the new terminal ileum at the ileostomy. They may be demonstrated by retrograde barium enemas, by lateral films of the abdominal wall with barium administration by mouth, and especially by ileoscopy with multiple biopsies. Although these are worth a trial with immunosuppressant agents to avoid continuing resection of the neoterminal ileum, they respond poorly to any except surgical procedures in my experience.

One extremely disturbing form of cutaneous fistulization that I have observed on several occasions is the rapid development of multiple cutaneous localized perforations of the anterior abdominal wall in patients with small bowel Crohn's disease—twice soon after the resection of active progressive inflammation, and a third some years following prior surgery of the small bowel Crohn's. They behave almost as if the underlying Crohn's was hyperactive and hyperaggressive.

One patient with progressive local resections which had failed to halt the progression and the eruption of new fistulas responded promptly to azathioprine so dramatically that the patient has refused any attempts at reduction of dosage or weaning from the drug during the past six years. One patient died before the immunosuppressant-agent era, and the third patient intolerant of 6-MP has so far responded to a course of cyclosporine. So perhaps my view regarding immunosuppressors should be revised in light of these hyperaggressive forms.

Labial Fistulas

A common clinical problem is the development of a localized inflammation of what is assumed to be the Bartholin glands of the vaginal labia of a patient with Crohn's disease. Almost invariably this local inflammation with tenderness and swelling is incised and drained only to continue to drain until it is part of a perivaginal fistula occurring in Crohn's disease of the rectum or anus. Control of the underlying intestinal disease usually leads to control and subsidence of the labial drainage.

Perirectal and Perianal Fistulas

Perianal granulomatous tissue, "Crohn tags," "elephant ears"—all these terms refer to those curious perirectal excrescences which are so characteristic of Crohn's disease that one can make the diagnosis by merely inspecting these tags. With further study, we usually invariably discover clear-cut evidence of Crohn's disease elsewhere in the intestinal tract. Usually tolerated by the patient as being only a nuisance, these tags may become quite troublesome and surround the rectum externally with what looks like a small cauliflower growth. I have so far seen only one histologic examination of these excrescences which contained no granuloma systems, and I have in general until fairly recently been reluctant to biopsy them or have then removed surgically, although a few patients and a few of my surgical colleagues are willing to attempt this. Thus far only one of my patients has had a debridement of the rectal area of such tissue, but follow-up of short duration already shows evidence of recurrence. I know that courses of azathioprine have done nothing for one patient suffering from huge Crohn's skin tags.

Perirectal and perianal fistulas were soon discovered to be part of the complex of Crohn's disease after the original description of the ileal form. It became clear early on that these did not arise from the ileal disease, but from the inflammation in the crypts of the anus. Almost invariably there was radiographic evidence of Crohn's elsewhere in the intestinal tract. Yet, in some rare instances, the localized form of rectal Crohn's disease may be the only evidence of the disorder.

Raised in an era when only incision and drainage of the perirectal abscess was considered and only if no other recourse was available in order to avoid the development of perirectal fistulas, my intention was always focused on the medical and surgical management of the "upstream" disease by whatever means were then considered useful. Actually, therefore, my stance in the past was to avoid surgery of these plaguing fistulas if at all possible and certainly to avoid fistulectomy. In recent years I have been convinced, however, that these true fistulas that do not respond to medical measures, should be treated by variations of the Park's procedure in which an intersphincteric abscess is dissected out and drained, especially when multiple fistulas exist. Certainly the one or two tolerated by a patient should be given a trial of antibiotics in the form of metronidazole and posssibly an immunosuppressant before concluding that surgery is the only alternative. It must be borne in mind that in this surgical procedure the external sphincteric muscle must be cut and the slight risk of incontinence is increased in those that have had prior incomplete or unsuccessful surgical attacks in this area or when previous abscesses have weakened both sphincters. My personal experience with long-term immunosuppressant agents in their curative role has been mixed; I have seen temporary improvement with azathioprine and relapse following withdrawal. Cyclosporine has healed one patient's stubborn persistent abscess and fistula after several fruitless surgical attacks and the use of 6-MP, and the patient and I are therefore reluctant to stop the cyclosporine for fear of severe relapse.

In men, perirectal fistula may begin to invade the scrotum and extend even up to the base of the penis and threaten the posterior urethra and epididymis. However, even in the presence of healed, or what appears to be healed, perirectal fistula, I have seen one persistently draining low-grade fistula extend high on the right side of the scrotum.

Most perirectal and perianal disease in Crohn's is easily diagnosed, but some individuals with perirectal and/or perianal pain and tenderness can present diagnostic problems. CT scans have increasingly been used and endosonography may help if the patient can tolerate the instrumentation. Magnetic resonance imaging (MRI) has been more recently advocated. My personal experience has been too limited to lead to any conclusions regarding their comparative values, but I intend to use all of these more frequently in the future. More frequent and puzzling diagnostically is the patient with a history of perianal disease, including fistulization who now complains of marked rectal pain and in whom physical examination as the first approach cannot be tolerated. I have found it useful to have the associated surgical colleague examine the patient under brief general anesthesia.

Other Rarer Forms of Fistulas

Crohn's disease of the esophagus is most rare indeed and its very existence debated. However, my attention has been called to at least two patients who had esophagotracheal fistulas that seem to have been in the Crohn's family rather than the more common perforation of hilar tubercular nodes into the esophagus. A diverting esophagogostomy was advised, but I have no information as to the subsequent fate of the patients.

Crohn's disease of the stomach and duodenum interestingly enough does not seem to fistulize to adjacent viscera. The one cutaneous duodenal fistula in Crohn's disease I have seen was of accidental origin. I have already recited the extremely rare cases of penetration of duodenal and jejunal diseases into the pancreas.

25

Pregnancy and Inflammatory Bowel Disease

Pregnancy raises interesting questions for the woman with IBD, some of which we can answer, and raises problems as well as joys for others. Almost as a rule, once the diagnosis of IBD has been confirmed for a young person, especially for a young woman, the second question that is almost invariably asked by patient, spouse, and parent is whether the patient will be able to have children.

In my experience, the answer in most instances is unequivocally yes, although I really believe that we lack solid information regarding the fertility rate of women with ulcerative colitis or Crohn's disease and the failure rate of the male with these disorders to impregnate his spouse. However, I do qualify my answer with the hedge that pregnancy in the female will occur provided ovulation has not been suppressed by the weight loss induced by the disease or steroid therapy and rarely in Crohn's disease by the tubal-ovarian involvement of direct extension from bowel disease. The only sex-linked disorder that I know associated with Crohn's is Turner's syndrome.

The afflicted male rarely fails to impregnate his spouse, except when excessively depleted by the illness or from the effects of sulfasalazine on his spermatogenesis. However we know correspondingly very little about the effects of IBD on this process in the presence of other medications. The suppression of the sperm count or motility due only to the use of sulfasalazine is apparently corrected by cessation of the drug within one to two months; fortunately, current evidence suggests that 5-ASA alone does not have the depressing effect of the 5-ASA sulfapyridine compound. But even a sulfasalazine induced reduction in sperm count cannot be considered common when judged by the complaints of the men in my practice.

Based on my impressions derived from a large and middle-class population, the ability of the female with Crohn's disease or ulcerative colitis to bear the fetus to term is no different from that in the general population except for the rare, extremely wasted patient. The rate of fetal wastage and miscarriage parallels the general rate of women of childbearing age.

When the fear of transmitting the disease to the offspring raises the question of this risk, reassuring answers are difficult to give. Family constellations occur in from 15 to 25 percent of the families of patients with IBD and there are no markers for predicting the risk for any one individual. There are some unpublished data that suggest that with one parent of a Jewish couple who has IBD, perhaps 10 percent of the offspring will inherit the disorder. In the case where both husband and wife have either of these disorders, solid information is even harder to come by, but available evidence certainly suggests that the risk is real and large, perhaps up to 35 percent, large enough beyond the realm of chance to make the question a sobering one. But I certainly would not counsel any afflicted couple against having children, though I would point out to them that the disorder is not sex-linked and it does not have a simple Mendelian pattern of inheritance.

What is the effect of pregnancy on the underlying disease? Pregnancy is not a disease, but it certainly brings with it physical and psychological stress of varying degrees often difficult to assess or predict. A considerable number of women insist that they have never felt better than during their pregnancies. One could rationalize this state of well-being as being due to the increase in endogenous steroid secretion from the placenta and the fetus's adrenals, but I do not know that this has ever been proven. What I am clear about is that, if the mother is in a state of remission or stability of her intestinal disorder, the pregnancy is likely to be a good one for her. If she is sick at the time of conception, the course is likely to be a more difficult one. This raises the question as to whether it would be advisable to limit attempts at conception by the mother. I would use the rule of thumb that a remission or stabilization of disease for a year would be advisable. However, in the woman who is growing older and determined to have a child if at all possible, my compromise would be six months of freedom from disease.

The more frequent problem that does arise is the question of whether the remission or stabilization is drug-dependent. I am well aware of the cumulative surveys which suggest that sulfasalazine and/or steroid therapy are safe during pregnancy and the degree of miscarriage or fetal wastage is not different from the general population. But still I do confess some anxiety regarding these drugs during the first three months of gestation. This anxiety is balanced by my experience that stopping these two classes of medication at the time of conception has led to exacerbation of the inflammatory bowel disease during this trimester in some patients. While there is every reason to believe that 5-ASA alone is probably a safer drug than sulfasalazine or steroids, I am aware of only one study of the effect of these newer 5-ASA drug products on the course of the mother's disease during pregnancy or fetal outcome, in which the drug was considered harmless.

If the mother's remission is drug-dependent on immunosuppressant drugs such as azathioprine or mercaptopurine, those with most experience with these compounds appear at present to have little fear of allowing the patient to remain on the drug if either inadvertently pregnant or even with a planned pregnancy. One senses that, if the patient is in remission on cyclosporine A, these workers would recommend the patient's pregnancy to go forward.

In my recollection of my experience, ileostomy, history of intestinal resection or colectomy or a cutaneous abdominal wall fistula or stable perineal disease should not interfere with pregnancy and vaginal delivery. Active perineal disease or draining fistulas or abscesses or rectovaginal fistula would lead me to advise cesarean section and to avoid episiotomy.

The issue of what intestinal imaging techniques are permissible during pregnancy is fairly clear in my own mind. I prohibit any x-ray exposure of a mother carrying a fetus. Clearly, this requires an obstetrical and menstrual history and a pregnancy test before suggesting a barium enema or GI series with small bowel films in any woman of childbearing age; yet I fear this is not always the custom of even the most careful clinicians. However, I have been advised that radiologists at present do ask these routine questions.

Inadvertent exposure to x-ray radiation of a female patient who is pregnant leads me to advise prompt interruption of the pregnancy, although I do not know that we have any scientific basis for this advice. The risk of inducing a spontaneous abortion or miscarriage in a pregnant woman with IBD by digital rectal examination, proctoscopy, or sigmoidoscopy is probably an acceptable risk if therapeutic management will be dependent on the examination and should be performed with care and minimal cleansing.

What about the treatment and management of IBD which has its onset or exacerbation during pregnancy? Certainly, ulcerative colitis or proctitis can occur at any time during the course of pregnancy and vary in intensity from the very mildest to the full-blown type. Clinical evidence and experience are clear that interruption of the pregnancy does not improve the patient's clinical course. First episodes of ulcerative colitis during pregnancy are rather rare in my experience, as well as those in the immediate postpartum period. Treatment of the sick, pregnant woman must be given as appropriately as if she were not pregnant. Topical steroids (foam preparations for minimal disease) and rectal hydrocortisone enemas (for more severe disease), together with or followed by oral steroids and/or azulfidine, are in order. Since the effects of 5-ASA in the newer form in pregnancy are not completely known, they are not suggested by me by the rectal route or by mouth. If local therapy does not work, then oral steroids or IV steroids in the form of hydrocortisone are clearly in order and are well tolerated. This viewpoint regarding 5-ASA needs to be balanced by a very recent report of its safety during pregnancy.

As to the therapeutic use of cyclosporine in pregnancy as therapy for the severely ill patient with ulcerative colitis facing colectomy after failure to respond to IV steroids, cyclosporine probably has no place. Since the incidence of severely ill patients with Crohn's disease or ulcerative colitis in pregnancy is rather rare in my experience, I can see little use for cyclosporine in this situation of uncertain evidence.

What is the situation regarding surgical intervention for ulcerative colitis, Crohn's disease, or intra-abdominal complications of the pregnancy? Appendectomy, abdominal laparotomy, and cholecystectomy for acute cholecystitis have been well tolerated by my patients and the pregnancy not interrupted. I would assume that early on in the pregnancy laparoscopic cholecystectomy could be performed. Appendectomy is also well tolerated, but appendicitis is often a difficult diagnosis to make. Unfortunately, for patients who are so sick as to require operation, subtotal or total colectomy is almost invariably followed by spontaneous abortion.

In a patient who has had Crohn's disease or ulcerative colitis flare-ups in the past, one must warn the patient that there is the possibility of a flare-up in the postpartum period. So I am prepared to treat them as soon as there are the first clinical signs of such a flare-up in the postpartum period. I do not think this episode should be managed in any way differently from the usual flare-up. It is also clear by now that, if the mother is healthy and in remission, one could consider the question of allowing her to nurse the child. It is generally felt at present that the use of 5-ASA, at least in the form of Azulfidine®, does not lead to Kernicterus or increases the physiological jaundice of the newborn. I am not aware of any evidence regarding the transmission of 5-ASA compounds in mother's milk. I think that if a patient requires steroids or immunosuppressants in the postpartum period they should not be allowed to nurse the child.

26

The Cancer Problem

Few clinicians doubt that the risk of developing cancer is real in patients with inflammatory bowel disease, especially ulcerative colitis, yet few of them have seen many such patients outside the larger tertiary referral centers in the United States and everyone accepts the two major risk factors: extent of disease and duration of disease. These are statistical findings. The problem remains: How is the individual clinician to proceed with the individual patient?

Cancer and Ulcerative Colitis

What are the actual risks? In our experience, the risk of cancer in universal ulcerative colitis compared to expected was twenty-five times the standard rate, whereas in left-sided ulcerative colitis it was seven times the expected, the risk of cancer not clearly higher than the expected rate in proctitis. However, it must be stressed that the rates stem from the era just before or during the early introduction of full colonoscopy. Further, I am always concerned with the finding of localized left-sided or rectosigmoid disease in an individual seen in past years to have had universal disease. In our earlier studies, all cancers occurred after two decades of universal ulcerative colitis and after three decades of left-sided disease. However, in my own and most clinical observers' experience, cancer in ulcerative colitis rarely occurred before eight to 10 years of illness, but is often very difficult to pinpoint accurately the date of the onset of the disease.

The fear that onset at an early age predisposes one to higher risk of cancer has been clearly seen as a question of the duration of the disease, not when it started. Indeed, duration has been so stressed in recent studies that the degree of clinical activity is said not to be important. But we have little secure data regarding the *number of years of clinical activity* that these patients have endured. All of us have seen colonic biopsies in patients with long-standing ulcerative colitis considered to be in full clinical remission which do show microscopic evidence of ongoing inflammation.

Pragmatically, I assume that patients with ulcerative colitis of 10 years' duration are at risk unless flexible sigmoidoscopy has shown that the colon above the rectosigmoid has been free of disease from the earliest clinical manifestations, and I accept an increasing cancer risk of 0.5 to 1 percent per year thereafter.

I will discuss the question of the surveillance of well patients shortly, but several clinical features of such individuals deserve emphasis at this point. Even though data are missing, a family history of cancer in first-degree relatives certainly raises my anxiety for a given patient. I would stress a further clinical point. Any symptoms suggesting a mechanical obstruction, however mild, raise the question of partial narrowing of the colon and/or the presence of a stricture. Living through the era of the introduction and routine use of total colonoscopy, I have seen a widespread change in attitude of clinicians to the problem of stricture. In the precolonoscopic era, a stricture in ulcerative colitis was automatically considered cancerous until proven otherwise by surgical removal. At present, our ability to pass even a small pediatric scope through the area with multiple biopsies being taken is considered adequate investigation.

A number of recent experiences of my own and of other experienced clinicians has shown me that this approach may be disastrous. In contrast to Crohn's disease and not completely so there, a stricture and a persistent narrowing of the colon in ulcerative colitis, no matter how well the patient is clinically, remains cancer in my opinion until full thickness pathologic study rules this out. (Obviously, any sessile lesion must be considered suspect as well.) Whether endoscopic sonography of narrowed areas which can be reached by current instrumentation or even CT scanning will help clarify this problem and reduce our concern remains to be demonstrated.

Colitis-Associated Cancers

Clinically, cancers of the colon in the presence of ulcerative colitis occur at an earlier age than the sporadic variety and often are multiple in number. The older impression that these colitis-associated cancers are more "malignant" than the ordinary cancer with a worse prognostic outcome has not been substantiated in recent experience.

Obviously cancer in ulcerative colitis presents no problem in clinical judgment. Patients who are sick and are failing to respond to conventional therapy of their preceding colitis, have little resistance to surgical intervention and colectomy, with the availability of rectal muscle-sparing operations, especially those with the complete mucosal stripping. However, in the presence of metastatic disease, one hesitates to add a permanent ileostomy to the patient's already difficult life situation and if an anastomotic operation preserving the rectum is possible it should be done. Even if no metastatic or regional lymph node spread is present, I am still reluctant to advise an ileorectal anastomosis, which contains any colonic mucosa, but frequent sigmoidoscopic examinations with biopsies remain a second and less desirable option.

Cancer Surveillance

The important and major problem is the surveillance of patients in relatively good health or with well-controlled illness on a modicum of medicines of acceptable risk and side effects who have had the disease for more than seven to ten years. Earlier suggestions regarding prophylactic colectomy did not strike a receptive ear of doctor or patient and is rarely heard at present. I do not suggest it and I doubt that many physicians have suggested it as well. Recognition of the concept of dysplasia as a forerunner of cancer and increasing agreement among experienced pathologists regarding the subtleties of grading the dysplasia has strengthened the clinical guidelines for the management of these patients in long-standing ulcerative colitis. Yet certain aspects still present clinical problems. Some patients with proven cancer may have no dysplasia, and some with high grade dysplasia may have no demonstrable cancers. Finally, only half of the patients with dysplasia have had cancer shown to be present at colectomy.

In the welter of these facts, how can or should we proceed? Colonoscopy of the entire colon to the ileocecal valve and multiple biopsied obviously must be done. If no mass, flat, or sessile lesion is seen, multiple orderly arranged biopsies throughout the entire colon need to be taken. I would go along with the timing of these colonoscopies at yearly intervals after the patient has had the illness for 10 or more years. However, we should remember this interval is an arbitrary one and is based on extrapolating the data of the doubling time of colonic metastases or missed lesions on x-ray. No one knows at what rate cancer doubles in size in ulcerative colitis!

If low grade dysplasia is found, I treat the patient with colitis vigorously and repeat the biopsies in three months. If we still find dysplasia, I consider colectomy, especially if any clinical features suggest the need for colectomy. The possibility of the continent ileostomy and the rectal-sparing operation certainly sugarcoats the bitter pill for otherwise healthy patients.

If dysplasia is associated with a local mass, the so-called DALM, then colectomy is clearly in order. A related similar problem of considerable vexation is discovery of a tubular villous adenoma in the presence of quiescent ulcerative colitis. Such an adenomatous polyp is by definition an example of high-grade dysplasia, but I am still reluctant to advise pancolectomy, especially if the polyp is in an area of well-healed colitis, but the possibility of rectal-sparing operations or ileorectal operations has led me to a more aggressive attitude recently.

Another not particularly rare situation at present is the question of surveillance of the rectosigmoid left in place following ileostomy in subtotal colectomy for ulcerative colitis. While the trend increases to remove the rectum in this type of colectomy or to do a mucosal stripping, and ileal pull-through, a considerable number of patients do have the rectum left behind. In my experience, sooner or later, all will have this disconnected segment removed by 10 years. Most male patients are reluctant, fearing impotence. My rule of thumb is to allow the patient to keep the rectum as long as the lumen can be visualized endoscopically and biopsied regularly. I do not rely on x-ray visualization. If this cannot be done at yearly intervals, proctectomy is my emphatic recommendation, especially if scarring or stricture prevents adequate visualization of this segment.

Cancer and Crohn's Disease

We have been slow in recognizing the threat of cancer in Crohn's disease, partially because it is a slowly developing complication and rarer than cancer in ulcerative colitis. Cancer in Crohn's disease is not common. It took 24 years in this institution, which is keenly interested in Crohn's disease, before the first cancer of the jejunum was recognized. However, our experience, biased as it may be coming from a tertiary referral center, indicates that cancer does occur more frequently, perhaps six to seven times more than would be expected by chance. Certainly carcinoma is not nearly as frequent in Crohn's disease as in ulcerative colitis. The facts in my experience are clear in this regard. The instance of cancer in the small bowel in patients with classic enteritis and ileocolitis is astronomically higher than the expected compared with specific cancer rates of standard populations. While much is made of the incidence of cancer with bypassed bowel as a factor, and my own experience is striking in this regard, I am beginning to believe that it is the long duration of the disease rather than the pure fact of the bypass that plays the important role. Cancer in IBD takes time to develop and in Crohn's disease it takes more time than in universal ulcerative colitis. It is known that in ulcerative colitis the extent of the disease, as well as the duration, is important. So it is not surprising that our overall experience in Crohn's is similar to that in our patients with segmental, left-sided ulcerative colitis.

The cancer in Crohn's disease in small and large bowel occurs mainly in grossly diseased tissue but not invariably so. Usually unicentric, in contrast to the multicentric cancers of ulcerative colitis, histology of the cancer in both disorders is very similar. Clinically, these neoplasms in Crohn's disease, which are not related to the age of onset, but to the entire duration of the disease, are difficult to diagnose before laparotomy and are associated with fistulas at times. (Cancer has been seen to arise in fistulous tracts on occasion.) The bypass loops are especially treacherous since they cannot often be visualized by barium enemas or a GI series. Cancer should be suspected in any patient who has had a bypass years before, certainly in anyone who after a long quiescent period suddenly has evidence of clinical activity, including a new fistula in the presence of a mass. The latter tendency has helped me on a few occasions to make the diagnosis before laparotomy. I believe we shall be grimly reaping the harvest of our bypasses of 20 to 30 years ago, and I am using any and all evidence of clinical activity to persuade patients to have the bypass loop removed. (Bypass loops done in infancy for congenital anomalies of the small bowel have been reported later in life to be associated with unusual malignancies.) I am persuaded that the overall small number of cancers of the colon which are the seat of Crohn's disease arises from the fact that most patients with Crohn's colitis do not tolerate this condition for ten or more years. We can't help but wonder what the long-term outcome as far as cancer is concerned will be in all those patients now being treated with vigorous immunosuppressant agents, not because of the drug, but because of patients' longer life with the disease.

Cancers Associated with Crohn's Disease

The interesting question that remains, however, is whether Crohn's disease anywhere in the GI tract is associated with a greater risk of cancer elsewhere. In other long-standing diseases of the small bowel, for example gluten enteropathy, the risk of upper gastrointestinal neoplasms and lymphoma of the GI tract has clearly been known for a long time to be increased over a control population. The point has not been established for Crohn's disease, although Adrian Greenstein and I have collected a series of patients with cancer of the colon in individuals who had Crohn's disease elsewhere in the small bowel. This may simply be a matter of chance in view of the increasing incidence of cancer of the colon in both men and women in recent decades, but one wonders what role the immunosuppressant therapies may play in the future in this regard and more information is needed regarding the incidence of microscopic Crohn's disease in the colons of these individuals. In most, small bowel Crohn's preceded the colonic cancer by years.

Associated Neoplasms in IBD

Midway between the general medical disorders associated with inflammatory bowel disease and the cancers of the intestinal tract in patients who have ulcerative colitis and Crohn's disease is the curious and intriguing question of other neoplasms that may bear some relation to their host's underlying inflammatory bowel disease.

Squamous cell cancer of the anus, which has been the seat of long-standing chronic inflammation, especially the perianal disease of Crohn's, seems easier to understand. These infrequent cancers can occur in the fistulous tract itself and seem clearly rated to the prolonged chronic inflammation.

Reticuloendothelial malignant lesions do occur in association with IBD and include lymphomas of the Hodgkin's variety and non-Hodgkin's variety, as well as leukemia.

Six cases of *acute myelogenous leukemia* in our experience at

Mount Sinai occurred in patients with long-standing ulcerative colitis and directed my attention to the possible role of the large amounts of diagnostic radiation they had had in the protracted course of the disorders. That they all fell into the group of premyelocytic leukemia suggest some homogeneous etiology, possibly x-ray radiation. Other cases in this category continue to be reported by other observers.

The association of *lymphoma* with IBD is equally intriguing and may parallel the association of intestinal lymphomas with long-standing gluten enteropathy. Be that as it may, our own series contained nine patients with lymphoma out of 1156 with ulcerative colitis and 1480 with Crohn's disease. In all four patients with Crohn's disease and in four of the five patients with ulcerative colitis the lymphoma was extraintestinal. Of these, two patients, one in each disease category, had also Hodgkin's; four had had lymphocytic lymphomas; and the remainder had large cell lymphomas (Crohn's disease one, ulcerative colitis two). Amyloidosis occurred in one of these patients with Crohn's disease. The clinical features of these patients' lymphomas are interesting because four with ulcerative colitis and two with Crohn's disease were asymptomatic with lymph adenopathy being the sole finding at the time of presentation. One of these had fever, night sweats, and weakness. Some had an abdominal mass or enlarging liver or spleen. The outlook was poor in general despite radiation and chemotherapy.

Malignant Melanoma

Although we have recently reported 10 patients with malignant melanomas of a large series of patients with inflammatory bowel disease (six with ileocolitis, one with regional enteritis, one with Crohn's colitis, and two with ulcerative colitis), it is not at all certain that this is a clear relationship. We have speculated that it may be related to immunosuppression from the diseases themselves, from our medical therapy, and/or from radiation. In these patients, the melanoma developed on the average 16 years after the onset of IBD.

27

Associated Diseases

The presence of two diseases presents interesting and intriguing problems. Are both manifestations of the same underlying cause? Or is one the consequence of the other? A discussion of some diseases associated with IBD seems appropriate before launching into the question of the extraintestinal manifestations of IBD.

It is clear that gallstones and kidney oxalate stones are frequent in Crohn's disease because of a now well-understood sequence of pathogenic events secondary to the intestinal disorder. However, the association of sclerosing cholangitis and carcinoma with sclerosing cholangitis in patients with ulcerative colitis is not easily related. Here the relationship of associated diseases and extraintestinal manifestations of IBD become blurred.

That some patients who have *psoriasis* and some with psoriatic arthritis also suffer from IBD, especially of the Crohn's variety, has been observed for a long time, but it has been emphasized recently as occurring more than by chance. Indeed, some HLA linkages have now been reported. Will this turn out to be analogous to the HLA

B27 association of ankylosing spondylitis and Crohn's disease? Our patients have told us for years that they had gallstones, and now we believe we know why. For years our patients told us that they had psoriasis, and now this association is well documented. In fact, some patients with IBD, especially Crohn's disease, have psoriasis and psoriatic arthritis analogous to the axial arthritis, ankylosing spondylitis, and sacroiliitis of Crohn's disease. Psoriasis linked with some HLA groups is not believed to be linked to HLA B27. The psoriasis itself does not fluctuate in intensity and activity with the underlying bowel disease and there is little evidence available at present regarding the distribution of the Crohn's disease in individuals with associated psoriasis.

Turner's Syndrome

On the basis of published reports, McConnell concluded that Turner's syndrome is associated more frequently with inflammatory bowel disease than could be encountered from chance which suggested to him the association is based on the abnormal state of the single X chromosome. This association is, however, very rare. The two occasions I thought I saw patients with Turner's and Crohn's disease turned out to be in reality only one instance: a young woman seen at 20-year intervals with classic Crohn's on both occasions.

Atopic Disorders

I have been impressed by the frequent association of IBD, especially Crohn's disease, with atopic disorders. Asthma, hay fever, and allergic rhinitis, often with family tendencies, have been frequent in my patients, and there is some literature that supports this impression.

Multiple Sclerosis

While an association has been suggested as occurring in multiple sclerosis and probably being linked in some fashion, this association must be extremely rare. I have only seen it once in a patient with Crohn's disease.

Myelodysplastic Syndromes

Although I discuss the problem of cancer of the gut in association with IBD illnesses in Chapter 26, it may be relevant to touch on the possibility that a series of patients my colleagues and I have seen with ulcerative colitis who developed acute myelogenous leukemia of the premyelocytic type did so as the result of their extensive exposure to radiation. The six patients from Mount Sinai all had the acute premyelocytic form of leukemia often associated with radiation, and others have reported similar cases. In addition, my colleagues and I have described some patients who have developed lymphoreticular neoplasms, as well, in inflammatory bowel disease. Recently the association of myelodysplastic syndromes with Crohn's disease has also been reported. These patients had refractory anemia with ring sideroblasts, and three of the patients were reported to have abnormalities of chromosome 20. One recent patient of mine had a myelodysplastic marrow, but the radiographic and clinical evidence of presumed Crohn's disease was evanescent.

Rare Metabolic Disorders with Intestinal Inflammation

The association of Crohn's disease, or rather Crohn's disease-like inflammatory lesions, with two rare metabolic entities raises interesting questions and possibilities in regard to the etiologic involvement of the function of neutrophils and macrophages.

Glycogen storage disease Type 1B has been reported in association with Crohn's disease or ileitis and is known to have neutropenia. This association known to pediatricians appears well established. Interesting to gastroenterologists is the recent report that two such subjects, 19 and 21 years old, improved their intestinal aspects remarkably when treated with granulocyte colony stimulating factor and/or granulocyte macrophage colony stimulating factor, presumably to improve the function of the neutrophils and/or macrophages.

The other metabolic disorder associated with Crohn's disease-like intestinal lesions is the Hermansky–Pudlak syndrome, a chromosomal recessively inherited syndrome with multisystem depositions of ceroid, albinism of eyes and skin, and bleeding tendencies. Neutrophils are normal in this order, but it has been hypothesized that the defect in the production of these lesions in these patients is the destruction of the function of the macrophage due to pigment accumulation.

Inflammatory bowel changes have been reported in other disorders with neutrophilic dysfunction, such as congenital neutropenia. Pancreatitis has also been associated with cyclic neutropenia, but neutrophilic function is not a primary defect in most cases of IBD. Yet various forms of intestinal enteropathy do occur in patients with neutropenia, some due to the CMV virus, or in neutropenic and immunocompromised patients, occasionally with clostridia, but not with more specific intestinal microorganisms, despite which antibiotics have generally been used. In this context also, the interaction of circulating leukocytes with the vascular endothelium, a subject of considerable research activity at present, is interesting from an etiopathogenic point of view, especially as regards Crohn's disease. Pounder and his group in London have recently emphasized the local importance of microvascular lesions in Crohn's disease, and others have observed the elevated levels of von Willebrand's factor in the blood in both Crohn's and ulcerative colitis as indicative of epithelial damage. The interaction of circulating leukocytes with the vascular epithelium can be elicited by platelet activating factor and the leukotriene LBT4 and, interestingly enough, platelet activating factor is inhibited by methotrexate. This will certainly lead to further interventional drugs in IBD.

Autoimmune Disorders Associated with Inflammatory Bowel Disease

Coexistence of some disorders generally held to fit the category of autoimmune disorders with IBD diseases has not only theoretical importance in the attempt to understand the etiopathology of both disorders, but more important present us with difficulties in managing the clinical conditions.

Most lists of these autoimmune diseases usually include autoimmune thyroid disease, hemolytic anemia, insulin-dependent diabetes, rheumatoid arthritis, systemic lupus, idiopathic thrombocytopenia purpura, and perhaps sclerosing cholangitis. When Snook and his colleagues at Oxford looked at the incidence of these disorders in IBD diseases compared with a controlled population, it was clearly increased in patients with ulcerative colitis (7 percent), compared with 2 percent in patients with Crohn's disease, and the control patient population. When sclerosing cholangitis was included, the percentage in patients with ulcerative colitis with autoimmune-associated diseases rose to 10 percent. Of these disorders in my own experience, the acute hemolytic anemia, and a form of cutaneous vasculitis with cryoproteins and cryoglobulins have been most interesting and also difficult to manage.

Acute Autoimmune Hemolytic Anemia

Are other associated diseases that we see occasionally occurring among large groups of IBD patients a matter of chance or treatment? In this context, the *acute autoimmune hemolytic anemia* that occurs infrequently in ulcerative colitis poses interesting questions of relatedness. I do not allude to the hemolytic anemia associated with treatment with sulfasalazine, which is ordinarily not associated with a positive Coomb's test, although this may occur occasionally, but to those anemias not related to treatment with this drug. In my three cases of autoimmune hemolytic anemia, one did not respond to medical management with steroids and azathioprine and required colectomy plus splenectomy; a second responded partially to splenectomy, and both hemolysis and colitis responded to azathioprine. Hemolysis in the third responded partially to steroids and completely to splenectomy. These three patients were all ill with their colitis, but not all patients with this type of hemolysis are. As with pyoderma gangrenosum, the hemolysis may appear years after a total colectomy.

28

Extraintestinal Manifestations

No small part of the clinician's fascination, as well as frustration, in treating IBDs arises from their extraintestinal manifestations—a bewildering array of disorders that are associated with the underlying disease and that may, at times overshadow the intestinal aspects or even appear long before we are cognizant of the IBD. To order my thoughts for the conceptual and therapeutic implications of these extraintestinal manifestations, I have modified a schema that my colleagues, Adrian Greenstein and David Sachar, and I had elaborated in 1976 and that I still find useful. These manifestations fall into four main broad groups:

1. One, which we have labeled as *colitis associated,* is closely related to the degree of clinical inflammatory activity of the bowel and appears to be correlated with the extent of intestinal involvement.
2. A second group of disorders we have labeled *pathophysiological consequences of intestinal disease.* These are the direct

consequences of disordered function and structure of the small bowel.
3. A third group I have provisionally thought of as being *immunologically mediated*.
4. The last group is composed of associated diseases, including extraintestinal neoplasms and some manifestations currently considered part of the autoimmune group.

The Colitic Group

In general this intriguing group of manifestations moves in keeping with the activity of the underlying bowel disease, being present or active when the bowel disease is most active and subsiding when the bowel quiets down, either spontaneously or as a result of therapies. Like most generalities, this one needs to be qualified at times. Further our own studies indicate that the colitic group occurs more frequently and is correlated with those patients who have large areas of the bowel involved, especially the colon, either in chronic ulcerative colitis or in Crohn's disease. The basic assumption is that they represent antigen–antibody immune complex disorders and deposition derived from the mucosa or absorbed through it. While there are some laboratory evidences of immune complex formation or consumption of complement factors, direct proof of immune complex deposition in the affected extraintestinal location is lacking. Rather, the concept is based on analogies with the arthritis of hepatitis B viremia or the manifestations of jejunoileal bypass operations for obesity. In the latter, immune complexes have been measured, and the antigens appear to be derived from intestinal bacteria. (Indeed, the enteritis of this complication of bypass operation bears interesting similarities to IBD.)

Eye Manifestations

Conjunctivitis, of either a diffuse type with injected conjunctival blood vessels or the phylyctenular variety (with tiny yellow nodules), is often seen just at the beginning of an episode of bowel inflammation or may often occur in the course of a clinical flareup. It is quite different from the chemosis or suffusion of the conjunctivae of patients on large doses of steroids. Anterior chamber involvement (iritis, choroiditis) occurs in my experience more frequently. This group of manifestations is rather closely tied to disease activity and responds to steroid suppression. More virulent eye involvement can occur including panophthalmitis, and I know of one patient with severe global ulcerations. These are extremely difficult to treat even with large doses of steroids. We might consider immunosuppressant drugs or even, in desperate cases, plasmapheresis in an effort to remove presumed immune complexes.

Articular Manifestations

Manifestations of arthralgias, fleeting and recurrent, as well as swelling, pain, and redness of peripheral joints, migratory in nature, and involvement of the axial skeleton (ankylosing spondylitis and sacroiliitis) are frequently present.

The peripheral forms should be clearly distinguished from the axial ones, since the peripheral joint manifestations clearly follow the course of the bowel disease activity and never lead to joint deformity of a permanent nature. On the one hand, ankylosing spondylitis and the often concomitant sacroiliitis on the other hand lead to fixed, stiff spines, and require management on their own, since they do not follow the bowel disease *paru-passu*. The well-documented association of HLA antigen B27 is related to the occurrence of the joint disease rather than to Crohn's disease with which it is associated. It is interesting, in this context, that the ankylosing spondylitis of *Yersinia enterocolitica* infection occurs primarily in HLA B27 individuals, highlighting a genetic susceptibility in this complication and suggesting analogies to Crohn's disease. Indeed we have recently seen a youngster with Crohn's disease and HLA B27 who himself is now beginning to manifest ankylosing spondylitis and sacroiliitis. His father and brother, both HLA B27 individuals, have ankylosing spondylitis yet no Crohn's disease. Unfortunately, treatment of the bowel disease has little effect on these axial arthridities.

We should also look at the problem of reactive arthritis from the viewpoint of the rheumatologist, especially in the light of arthritis associated with enteric infections. In addition to *Yersinia, Campylobacter jejuni, Salmonella* infections, as well as *Shigella*, can activate a polyarticular, migratory joint involvement. All these occur in association with the HLA antigen B27, paralleling the same HLA linkage with Crohn's ankylosing spondylitis.

Some Difficult Problems in Joint Manifestations

The peripheral joint manifestations, the "reactive group" closely related to the activity of the underlying inflammatory bowel disease and the more stubborn axial group (including ankylosing spondylitis and sacroiliitis), whose course is not tied to the underlying disorder, are usually easily recognized and diagnosed, although the latter group is much more difficult to manage and treat.

More disturbing are some examples of peripheral joint symptoms and manifestations, which are harder to classify. I have already mentioned stiffness of joints and diffuse aches following withdrawal from steroids as well as the pseudorheumatism of steroid use. Time usually clarifies these annoying problems.

More difficult is the patient with peripheral joint manifestations that resemble rheumatoid arthritis, but are sero-negative for markers of that joint disease. Some also lack any evidence of joint erosions on x-ray, further separating them from being easily classified as fitting into the rheumatoid group. When the underlying inflammatory bowel disease response to current medical therapy and then their associated joint symptom complaints also subside, the question of classification becomes moot and/or academic. It is when their intestinal and joint manifestations are slow to respond that classification and more important management becomes difficult. One such example is a young male with a long history of ulcerative colitis who continues to complain of severe disabling axial joint symptoms and swelling of hands and feet that are sero-negative and whose joints reveal no erosions, although his hands look like rheumatoid hands. Of all the drugs exhibited (5-ASA in oral and rectal forms in all current preparations, azathioprine, hydroxychloroquine, methotrexate, and steroids) only long continued dosages of oral prednisone suppressed his local intestinal and joint manifestations. One certainly hesitates to suggest colectomy in such an instance even as part of a rectal-saving operation. While for some individuals pyoderma gangrenosum has been the sole indication for colectomy in a selected limited number of cases with generally excellent results, in some, months have been required for complete healing, I have as yet not advised pancolectomy for peripheral joint manifestations alone, although I am prepared to use cyclosporine by the IV route before surgical intervention.

Skin Manifestations

While *erythema multiforme* occurs in IBD it must be exceedingly rare, much more common and clinically important are erythema nodosum and pyoderma gangrenosum. In my experience, the former's more likely in Crohn's disease, the latter in ulcerative colitis.

Erythema nodosum, assumed to be a manifestation of antigen–antibody complex deposition, but not as far as I am aware ever proven to be, is a very good index of bowel inflammatory activity and usually responds quite promptly to treatment of the underlying disease especially with steroid therapy. Biopsies of erythema nodosum should be avoided. In some cases necrosis and ulceration of the nodosum may occur, but it usually heals completely, sometimes with extensive scarring.

Pyoderma gangrenosum, whose etiology is equally obscure, does contain an element of vasculitis on biopsy, and most frequently occurs in the lower extremities. Like erythema nodosum, pyoderma gangrenosum usually occurs in the presence of an active bowel disease, but not always, and most frequently it resolves when the bowel disease does. Pyoderma gangrenosum can be most painful and threatens the structure of the foot's long tendons. Local steroid therapy may make the patient more comfortable, but does not usually solve the problem. In patients with active disease, steroid therapy is the method of choice, but may have to be reinforced with other immunosuppressant drugs in the stubborn intractable case. Sulfonamides of the Dapsone variety are often tried, but I have not found this drug particularly helpful.

Pyoderma gangrenosum can be very stubborn, and colectomy or resection of involved bowel, usually the colon, is often recommended in these instances. Does operation cure pyoderma gangrenosum? Not always. In our experience, colectomy helps dramatically in those with active disease of a severe variety and healing may occur often even before the patient leaves the hospital. In those with mild to moderate activity, colectomy helps, but healing is much slower, however, all our patients healed within a year of the operation with the aid of other medications.

We have also learned that pyoderma may occur years before recognition of the underlying bowel disease and, more frustratingly, that it may occur, or reoccur, following colectomy. So I have learned to be quite cautious in discussing its prognosis with the patient before advising colectomy.

In addition to the anomalies of response to treatment, pyoderma gangrenosum may occur in the presence of an inactive bowel disease. In several such instances, the skin lesion occurred at the sites of trauma to the extremities involved and we have observed pyoderma of the hands and arms at the site of injury. Steroid therapy seems to be effective in these instances as well, facts that dermatologists have apparently long known.

Care must be taken to differentiate the lesions of pyoderma from other related cutaneous lesions in Crohn's disease. Equally interesting is the lesion of *metastatic Crohn's disease*. The presence of a clinical classic Crohn's lesion as a result of spillover from the gut into the larynx has been well documented. Cutaneous granulomatous lesions occurring near skin fistulas, either in the abdominal wall or in the perineum, are equally well known. But by metastatic Crohn's disease, we mean a skin lesion in no relation to the gut or to a fistula arising from the gut. Usually occurring in intertrigenous areas in the groin in men, beneath the breasts in women, seen by me in the cheek as well, the swollen and the ulcerated lesion contains the characteristic hard sarcoid granulomatous nodules and responds to standard steroid therapy. Metastatic Crohn's disease certainly raises interesting possibilities of the disseminated nature of the disease. Other such granulomatous nodules have been seen in the liver of course and even in the bone marrow.

The Pathophysiologic Group

This variety of extraintestinal manifestations have as their basis disturbances in the physiology of the small intestine and hence arise mainly in Crohn's disease involving the distal small bowel. Since the ileum is central in Vitamin B12 and fat absorption and bile-salt reabsorption, distortions of function related to these substances comprise this category.

Malabsorption Syndromes

VITAMIN B12 MALABSORPTION

Vitamin B12 malabsorption due to impaired ileal absorption of the vitamin-intrinsic factor–calcium complex is rarely complete in Crohn's ileitis except after multiple resections have removed all or almost all of the ileum. Partial failure of Vitamin B12 absorption is common and constitutes and contributes to the complex anemia of the patient.

FAT MALABSORPTION

Fat malabsorption with its attendant weight loss in those well enough to eat an adequate diet follows of course from the impaired micellar lipid formation secondary to loss of bile salts in the stool as a result of impaired or complete loss of ileal reabsorptive surface. This loss of fat is rarely total and can be compensated by the mass action of increasing dietary intake or the use of medium chain triglyceride supplements, but the capricious appetite of these sick patients makes this approach difficult. The malabsorption of fat may be due also to loss of absorptive surface in those so unfortunate as to have so many resections as to result in a short small bowel.

Gallstone Formation and Biliary Tract Disease

Older clinicians have known for a long time that the incidence of gallstones was high in Crohn's disease. Now the elucidation of the role that the loss of bile salts and consequently reduced bile-salt pool play in biliary colic and calculi formation has made this explicable. Except in the case of classic biliary colic, it is difficult at times to sort out the distress due to biliary tract disease from their intestinal discomfort in these patients, especially when the right colon is involved in the disease. A large number of patients require cholecystectomy and thus add to their bile-salt induced catharsis via the colon. Cholestyramine, helpful to some patients with bile-salt catharsis, may aggravate their fat malabsorption.

Kidney Aspects

URIC ACID STONES

Renal colic and hematuria due to uric acid stones or passage of uric acid crystals in the urine are commonplaces in IBD, especially Crohn's, and frequently in patients who have had an ileostomy for Crohn's disease or ulcerative colitis who are prone to dehydration. At times women, as well as men, may have an elevated level of uric acid in the blood not related to gout, but I have always presumed that this was due to release of tissue breakdown products secondary to the inflammatory process. Loss of intestinal fluid secondary to diarrhea or in those patients with an ileostomy higher than customary leads to impaired perfusion of the kidneys. The loss of base in the stool often makes it extremely difficult to alkalinize these patients' urine, and they may be plagued with persistent passage of renal uric acid stones.

OXALATE STONES

The elucidation of the mechanism by which these stones arise secondary to hyperoxaluria has been a brilliant contribution to our understanding of this complication of ileal disease, with the loss of ileal function in the presence of a functioning colon. Fat malabsorption leads to loss of calcium in the stool with increased attendant reabsorption of oxalate and secondary hyperoxaluria. The therapeutic maneuvers directed against this state include a low-oxalate, low-fat diet; removal of large doses of Vitamin C from the diet; adequate intake of fluid; reduction of concomitant hyperuricemia by xanthine oxidase inhibitors; and addition of calcium to the diet.

OBSTRUCTIVE KIDNEY DISEASE:
HYDROURETER AND HYDRONEPHROSIS

Hydroureter and hydronephrosis occur as a direct result of extension of the intestinal inflammatory process from the ileum and right colon in most instances and rarely bilaterally from Crohn's disease of the whole colon. This periureteral inflammation rarely causes renal symptoms and does not affect the urine. Often silent and diagnosed only by routine IVP or more often sonographically these days, those patients with this syndrome usually have psoas spasm, a limp or pain in the upper right thigh. The important point is that the physician should make every effort to avoid any surgical attack on the kidney. Ureterallysis in this situation in contrast to the hydronephrosis and hydroureter of retroperitoneal fibrosis is rarely, if ever, required. Treatment of the underlying disease, whether by medical or surgical attack, has resolved this problem in my experience.

Parenthetically I want to call attention to the curious symptom of an atonic bladder seen more frequently in women, but occasionally in men, and those with acute colonic disease of a fulminant and severe form, which I have ascribed to involvement of the pelvic nerves. On one occasion, this was the chief complaint of women with acute ulcerative colitis of unusual distribution.

Fistulas to the urinary bladder are discussed in Chapter 24 on fistulas where dysuria, pneumaturia, and microscopic hematuria call attention to the extension into the bladder of the disease of the bowel in Crohn's disease. Gross hematuria is rare but can be a frightening experience as when, in one instance, the urologist in the course of cystoscopy observed a blood vessel in the wall of the bladder spurting away.

The Immunologic Group

Amyloid in IBD

Although amyloidosis is a rare complication of extraintestinal manifestation of IBD, we have seen 25 patients in 40 years at Mount Sinai among 3000 cases of IBD. Since the original description of this extraintestinal manifestation from this institution it soon became clear that amyloid was associated usually with Crohn's disease. Only three patients with ulcerative colitis had amyloid of those seen by our group, two of whom by me.

Amyloid may be considered as a metabolic disorder of IBD, but the mechanism is not understood except perhaps in some way as being related to inflammation, but not necessarily to suppuration, possibly by way of stimulation of macrophages. Currently the majority of patients with amyloid are males in contrast to the equal sex distribution of Crohn's disease; it takes many years to develop clinically obvious amyloid if we calculate from the onset of the clinical manifestations of the disease.

Our patients were diagnosed by renal, intestinal, colonic, and thyroid biopsies during life; from the histological examination of resected bowel; and at postmortem. We did not use biopsies of the abdominal fat pad, a mechanism now advised by the leading investigators in this field. Amyloid A-protein was then demonstrated by tissue staining, but we have had only a few measurements in this study of serum amyloid A-protein. This marker should be used more frequently in the future. In many, amyloid was found in multiple sites and the majority had amyloid in more than one site.

In any consideration of pathophysiology, the associations of amyloid with other extraintestinal manifestations are of considerable interest and occurred in more than half of these subjects with Crohn's disease, as well as those with ulcerative colitis. The "colitic manifestations" include peripheral arthritis and sacroiliitis, erythema nodosum, pyoderma gangrenosum, and aphthous stomatitis. Of the small bowel pathophysiologic associations, there was one patient with kidney stones but without obstruction or infection and four instances of gallstones. Suppurative conditions were present in many of our patients with Crohn's disease and amyloid. These included intestinal and external fistulas, perianal abscesses and fistulas, and intra-abdominal abscesses, including a few individuals with liver abscess. Only one patient developed an incidental retroperitoneal lymphosarcoma diagnosed at the time of the appearance of the amyloid.

CLINICAL PRESENTATION OF AMYLOID

We have observed six clinical patterns in patients with IBD. First, renal amyloid present in both Crohn's disease and ulcerative colitis with mild or massive proteinuria and/or renal insufficiency in 20 patients. Second, severe malabsorption due to small bowel amyloid was the presenting feature in two subjects. Third, amyloid cardiomegaly occurred in a young man of 31 with congestive heart, arrythmia, and generalized anasarca following subtotal colectomy for Crohn's disease. Fourth, two of the very early cases who came to postmortem had hepatosplenolomegaly due to the infiltration of liver and spleen. Fifth, three patients with renal amyloid had major amyloid deposits in the thyroid and one deposit so large that partial thyroidectomy was required. Sixth, in four subjects coming to autopsy, there was generalized amyloidosis with liver, kidney, and pancreas in all; bone marrow, spleen, and GI tract in three; and thyroid, heart, ovary, and adrenal in two.

None of the 17 subjects had a family history of Familial Mediterranean Fever, although many were Jewish. The seriousness of amyloid as a complication is underlined by the fact that half of the CD patients died; nine of the 17 who died had amyloid associated with renal failure. No evidence was forthcoming from this series that resection of these bowels resulted in amelioration or cure of amyloid disease. There are few patients reported in the literature who improved following surgery, although a rare case probably does exist.

TREATMENT OF RENAL AMYLOID

The older orthopedic literature reports the resolution of amyloid following surgical treatment of the amyloid associated with chronic osteomyelitis, but there is no credible report on the amyloid of IBD responding to any form of medical therapy, and steroids have been implicated in some experimental studies in the pathogenesis of amyloid. Following the discovery that colchicine not only could abort episodes of Familial Mediterranean Fever, but prevent its appearance, prevent its associated amyloid, and even lead to the resolution of amyloid, it seemed logical to me to study the effects of colchicine on the renal amyloidosis of IBD. The two patients with massive proteinuria and renal amyloid who received 0.6 mg of colchicine twice daily had a remarkable response and have been reported separately. Both had ulcerative colitis. One, a female of 24, excreted 9 grams of protein in 24 hours in her urine, and had a reduction in the 24-hour urine to 5.1 grams in three months and to 0.5 grams in eight months with marked improvement in creatinine clearance. The second, also a female, with urinary protein excretion of 13.7 grams in 24 hours, had a reduction in the 24-hour period of excretion to 6 grams in two months and gradual reduction in the next nine years of continued therapy to 0.37 grams in 24 hours. Both patients tolerated this drug very well, and it induced no diarrhea. The third patient, a male with CD with marked renal failure and marked proteinuria, went on to peritoneal and hemodialysis while awaiting transplantation. Although this patient took the same doses of colchicine for only a few weeks and had a marked reduction in serum amyloid A-protein, he went on to die after rejecting two kidney transplants within a short period of time. The survival and improvement in the two other patients was and remains impressive, and I believe that colchicine should be used as soon as a diagnosis is made of amyloid by blood or tissue evidences and continued indefinitely. Just how colchicine in these instances works is unknown, but colchicine binds to intratubular structures and has been shown by several groups of workers to inhibit interferon gamma expression of HLA-DR on gut epithelial cell lines. A trial with colchicine in ulcerative colitis and Crohn's disease certainly seems worth doing even without amyloid.

Associated Diseases

Liver Involvement

It can easily be argued that involvement of the liver is the commonest extraintestinal manifestation of inflammatory bowel disease as widely varying but substantial instances of hepatic abnormalities are reported in the literature in these disorders. Some, like primary sclerosing cholangitis (PSC) and cholangiocarcinoma, seem to be associated with ulcerative colitis; gallstones, granulomas, and amyloid are more frequently associated with Crohn's disease. The most severe of these complications—PSC and cholangiocarcinoma—curiously enough are not related to the clinical activity of the associated colonic disease.

While I once assumed along with my clinical colleagues that minor abnormalities of liver function tests such as the SGOT, SGPT, and alkaline phosphatase levels of the serum were the innocuous manifestations of "pericholangitis," I have been convinced by others' as well as with my own experience that these early laboratory studies are indeed of more than minor significance and point to the steady evolution of PSC. I now pay more attention to these findings than I did and more likely have liver biopsies done in asymptomatic individuals, as well as resorting to endoscopic retrograde cholangiopancreatography (ERCP) in those with the earliest clinical or laboratory signs of cholestasis. I have learned that colectomy for ulcerative colitis confers no benefit on the associated PSC. I heed the warning of other observers that the formation of an ileostomy stoma may lead to very difficult to control persistent varices and gross bleeding.

I have seen over the years the list of ineffectual drugs for PSC grow to include corticosteroids, azathioprine, antibiotics, penicillin, colchicine, methotrexate, and cyclosporine, even though some of these agents reduce the accompanying colitic activity. Current trials of the bile salt ursodecholic acid seem attractive in view of the bile salt's low toxicity and some reports of improved liver function tests. I await the long-term results of current trials of methotrexate, fearful of its known hepatotoxicity, but of course with larger doses in the past.

Liver transplantation is the reasonable option for patients with increasing disability and inevitable risk of liver failure. This is never an easy decision to help the patient make, but one should not delay because the colitic activity is stable in the vain hope that the patient somehow will also improve. The sudden development of acute flare-up of the underlying colitis has in several of my patients prevented the long planned for transplantation and ended with disastrous results.

Vascular Complications

Vascular complications of IBD are interesting and sometimes important. I have already enumerated in the section on laboratory studies in Chapter 7 that a marked increase in the number of platelets of these patients' disorders is often seen and usually considered as an acute phase reactant. These are functional platelets as already indicated and they are often used to explain vascular complications or as contributing to the hypercoagulable state associated in these patients. This is a convenient place to point out again that a newer hypothesis for the pathogenesis of Crohn's disease recently advanced by Pounder and his colleagues in London is that the partial and patchy distribution and some of the pathologic changes of that disease are due to vascular damage of smaller blood vessels of the intestine and colon. Consistent with this and also of possible importance in the vascular complications is the growing literature of the possible role of Factor VIII (von Willebrand's factor [VWF]). VWF levels in the blood appear from some preliminary studies to be increased in ulcerative colitis, infectious colitis of bacterial origin, and in Crohn's disease. In the first two instances, this seems to be related to the consequences of inflammation in active disease as an acute phase reactant. The elevated levels subside after treatment or improvement. Their presence in Crohn's disease may be more important since the blood levels of VBF, a very useful marker of vascular damage, remains elevated even during remission or inactivity of the disease, suggesting some possible fundamental role in the pathogenesis of the disorder and certainly able to contribute to the onset of pathologically significant vascular complications.

Variants of acute vascular disorders have been reported in IBD, some even in the retina. My colleagues and I have reported several cases of patients with cyanosis and gangrene of the fingers and toes in Crohn's disease associated with an increase in cryoglobulins and in cryofibrinogenemia, which responded to steroid and anticoagulants. I have always viewed ileofemoral venous thrombosis, seen more frequently in ulcerative colitis, although of lesser incidence in recent years, as having an extremely grave prognosis for these unfortunate, often younger, individuals, yet pulmonary embolization does not seem to occur in patients with IBD except rarely as a postoperative phenomenon.

Tragic but fortunately quite rare is the occurrence of cerebrovascular accidents with attendant hemiplegia in young adults. Several of these instances in my practice, and reported by my neurological colleagues, have occurred both with and without steroids and without other evidence of a hypercoagulable state or diabetes. Often they are superclinoidal in location. Postmortem review in a few instances of this kind did not reveal significant atherosclerosis of visible vessels.

Extraintestinal Manifestations of Medical Therapy

Some of the extraintestinal manifestations of our drug therapy, their classical side effects, are easily separated from the manifestations of the underlying inflammatory bowel disease but others are hard to classify.

Of those associated with the 5-ASA group of medicines, fever alone, peripheral arthralgias, nausea and depression of appetite, abdominal pain secondary to pancreatitis, and diarrhea, often seen with some variants of 5-ASA but emphasized in my experience with olsalazine, may mimic the disease we are attempting to treat. The steroid complications are well known and usually well recognized, but some are difficult to separate from the underlying disorder. Early aseptic necrosis of bone may take time to declare itself fully and may appear simply as the arthralgia of IBD. During the period when we attempt to wean our patients from steroids, muscle aches and stiffness, joint discomfort, lassitude and weakness, and even the abdominal cramping of the withdrawal syndrome may press us hard to separate these side effects from a possible flare-up of the basic disorder.

The side effects of the immunosuppressant group of medicines, azathrioprine and 6-mercaptopurine, usually present no real difficulties but occasional abdominal pain with its pancreatic inflammatory effects may mimic the biliary pancreatitis of Crohn's disease and the colitic tenderness of ulcerative colitis.

Extraintestinal Manifestations of Surgical Therapy

The small amount of active ulcerative colitis mucosa of the rectal segment left behind in a Hartmann pouch in the first-stage operation of subtotal colectomy, for example, may continue to maintain some of the extraintestinal manifestations with which the patient first presented, pyoderma gangrenosum being the most troublesome in my experience. *Diversionary colitis* in an excluded segment of colon after some form of operation for Crohn's disease has not often shown itself in my experience with extraintestinal manifestations. While pouchitis of a severe persistent form seems to occur more frequently in individuals with many preoperative extraintestinal manifestations, it does not itself in my experience present with many of these symptoms when it does develop. The more common and more easily treated form of pouchitis responds to metronidazole therapy and seems to show little extraintestinal manifestations.

Factitious Extraintestinal Manifestations

It is no surprise that healthy individuals may present to their physicians with factitious disease. Gastroenterologists are saddened but not surprised when some of their patients with secretory diarrhea have been surreptitiously ingesting laxatives. What is more puzzling is the patient with known inflammatory disease who manufactures extraintestinal manifestations. My main examples had classical Crohn's disease of small and large bowel who presented themselves with recurrent ulcerations in the anterior chest wall and abdomen. These deceptions were revealed when one patient fell asleep with her sharp cuticle nail scissors imbedded in her skin or others were observed digging into their skin with a sharp scissor. Another patient with severe proctitis who required a diverting sigmoid colostomy had induced recurrent ulceration in the colostomy with a safety pin.

Finally, it should be remembered that some laxatives can mimic for a while ulcerative inflammatory disease. The prolonged use of an old-fashioned laxative, Hinkle's pills, resulted in one instance in mild changes in small bowel and colonic x-ray films in the preendoscopic era with the passage of traces of blood in the stool and the passage of occult blood.

A Personal Bibliography

Chapter 1. Introduction

Crohn BB, Janowitz HD: Reflections on regional ileitis twenty years later. *JAMA* 1954;156:13–22.

Sachar DB, Janowitz HD: Inflammatory bowel disease. *Disease-A-Month*, July 1974;1–44.

Janowitz HD, Sachar DB: New observations in Crohn's disease. *Ann Rev Med* 1976;27:269–285.

Janowitz HD: Crohn's disease: 50 years later. *N Engl J Med* 1981;304:1600–1602.

Janowitz HD, Sacher DB: Inflammatory bowel disease. *Adv Int Med*, 1982;205–246.

Chapter 2. Disease or Diseases?

Lindner AE, Marshak RH, Wolf BS, Janowitz HD: Granulomatous colitis: A clinical study. *N Engl J Med* 1963;269:379–385.

Janowitz DH, Lindner AE, Marshak RH: Granulomatous colitis: Crohn's disease of the colon. *JAMA* 1965;191:825–828.

Present DH, Lindner AE, Janowitz HD: Granulomatous diseases of the gastrointestinal tract. *Ann Rev Med* 1966;17:243–256.

Chapter 3. The Nature of the Beasts

Janowitz HD, Present DH: Granulomatous colitis: Pathogenetic concepts. *Gastroenterology* 1966;51:778–784.

Present DH, Chapman ML, Cohen N, Janowitz HD: The correlation of sigmoidoscopy and rectal valve biopsy in granulomatous disease of the small bowel. *Gastroenterology* 1967;52:1113.

Chapter 4. Theories of Etiology and Their Therapeutic Implications

Transmission Studies

Sachar DB, Taub RN, Janowitz HD: A transmissible agent in Crohn's disease? New pursuit of an old concept. *N Engl J Med* 1975;293:354–355.

Taub RN, Sachar DB, Janowitz HD, et al: Induction of granulomas in mice by inoculation with tissue homogenates from patients with inflammatory bowel disease and sarcoidosis. *Ann NY Acad Sci* 1976;278:560–564.

Taub RN, Sachar DB, Siltzbach LE, Janowitz HD: Transmission of ileitis and sarcoid granulomas to mice. *Trans Assoc Am Phys* 1974;87:219–224.

Cellular Immunity

Meyers, S, Sachar DB, Taub RN, Janowitz HD: Anergy to dinitrochlorobenzene and depression of T lymphocytes in Crohn's disease and ulcerative colitis. *Gut* 1976;17:911–915.

Meyers S, Sachar DB, Taub RN, Janowitz HD: Significance of anergy to dinitrochlorobenzene (DNCB) in inflammatory bowel disease: Family and postoperative studies. *Gut* 1978; 19:249–252.

Sachar DB, Taub RN, Brown SM, Present DH, Korelitz BI, Janowitz HD: Impaired lymphocyte responsiveness in inflammatory bowel disease. *Gastroenterology* 1973;64:203–209.

Ramachander K, Sachar DB, Janowitz HD, Forman SP, Douglas SK, Taub RN: B lymphocytes in inflammatory bowel disease. *The Lancet* 1974;2:45–46.

Sachar DB, Taub RN, Ramachandar K, Meyers S, Forman SP, Douglas SD, Janowitz HD: T and B lymphocytes and cutaneous anergy in inflammatory bowel disease. *Ann NY Acad Sci* 1976;278:565–573.

Godin NJ, Sachar DB, Winchester R, Simon C, Janowitz HD: Loss of suppressor T cells in active inflammatory bowel disease. *Gut* 1984;25:743–747.

EB Virus in IBD

Grotsky HW, Hirshaut Y, Sorokin C, Sachar D, Janowitz HD, Glade PR: Epstein–Barr virus and inflammatory bowel disease. *Separatum Experientia* 1971;27:1474–1475.

C. difficile Toin

Meyers S, Mayer L, Bottone E, Desmond E, Janowitz HD: Occurrence of *Clostridium difficile* toxin during the course of inflammatory bowel disease. *Gastroenterology* 1981;80:697–700.

Chapter 5. Modes of Clinical Presentation

Ulcerative Colitis

Janowitz HD: Chronic inflammatory disease of the intestine. In Beeson PB, McDermott W, Wyngaarden JB (eds): *Cecil Textbook of Medicine*, ed 15. Philadelphia, WB Saunders, 1979;1560–1578.

Crohn's Disease

Janowitz HD: Chronic inflammatory disease of the intestine. In Beeson PB, McDermott W, Wyngaarden JB (eds): *Cecil Textbook of Medicine*, ed 15. Philadelphia, WB Saunders, 1979;1560–1578.

Present DH, Lindner AE, Janowitz HD: Granulomatous diseases of the gastrointestinal tract. *Ann Rev Med* 1966;17:243–256.

Marshak RH, Lindner AE, Janowitz HD: Granulomatous ileocolitis. *Gut* 1966;7:258–264.

Marshak RH, Janowitz HD, Present DH: Granulomatous colitis in association with diverticula. *N Engl J Med* 1970;283:1080–1084.

Janowitz HD, Sachar DB: Clinical, laboratory, and certain differential diagnostic features of noncolitic Crohn's disease. In Kirsner JL, Shorter RG (eds): *Inflammatory Bowel Disease*, ed 2. Philadelphia, Lea & Febiger, 1980, pp 150–165.

Meyers S, Janowitz HD: Crohn's disease, clinical features. In Berk JE, Haubrich WS, Kalser MH, Roth JLA, Schaffner F (eds): *Bockus Gastroenterology*. Philadelphia, WB Saunders, 1985, pp 2240–2259.

Children

Janowitz HD: Little people, big people, and Crohn's disease. *Gastroenterology* 1978;74:951–952.

Clinical Varieties of Crohn's Disease

Greenstein AJ, Lachman P, Sachar DB, Springhorn J, Heiman T, Janowitz HD: Perforating and non-perforating indications for repeated operations in Crohn's disease: Evidence for two clinical forms. *Gut* 1988;29:588–592.

Chapters 9–11. Difficult and Differential Diagnoses

Wang C, Janowitz HD, Adlersberg D: Intestinal lipodystrophy (Whipple's disease) amenable to corticosteroid therapy. *Gastroenterology* 1956;30:475–488.

Kogan E, Janowitz HD: Intestinal tuberculosis: Difficulties in diagnosis in the absence of florid pulmonary involvement. *J Mt Sinai Hosp* 1956;23:597–615.

Cohen N, Paley D, Janowitz HD: Acquired hypogammaglobulinemia and sprue: Report of a case and review of the literature. *J Mt Sinai Hosp* 1961;28:421–427.

Marshak RH, Wolf BS, Cohen N, Janowitz HD: Protein-losing disorders of the gastrointestinal tract: Roentgen features. *Radiology* 1961;10:225–230.

Marshak RH, Janowitz HD, Present DH: Granulomatous colitis in association with diverticula. *N Engl J Med* 1970;283:1080–1084.

Siltzbach LE, Vieira LOBD, Topilsky M, Janowitz HD: Is there Kveim responsiveness in Crohn's disease? *The Lancet* 1971; 8:634–636.

Carcinoembryonic Antigen

Rule AH, Straus E, Vandervoorde J, Janowitz HD: Tumor-associated (CEA-reacting) antigen in patients with inflammatory bowel disease. *N Engl J Med* 1972;287:24–26.

Rule AH, Goleski-Reilly C, Sacher DB, Vandevoorde J, Janowitz HD: Circulating carcinoembryonic antigen (CEA); Relationship to clinical status of patients with inflammatory bowel disease. *Gut* 1973;14:880–884.

Chapter 12. Assessing the Patient's Clinical Status

Korelitz BI, Janowitz HD: The physiology of intestinal absorption. *J Mt Sinai Hosp* 1957;24:181–205.

Kogan E, Schapira A, Janowitz HD, Adlersberg D: Malabsorption following extensive small intestinal resection including inadvertent gastroileostomy. *J Mt Sinai Hosp* 1957;95:320–325.

Gerson CD, Cohen N, Janowitz HD: Small intestinal absorptive function in regional enteritis. *Gastroenterology* 1973;64:907–912.

Meyers S, Lichtiger S, Feuer EJ, Lachman EA, Janowitz HD: Fecal alpha-1-antitrypsin as a measure of Crohn's disease activity: The effect of therapy and anatomical extent of disease. *J Clin Gastro* 1988;10:491–497.

Chapter 13. Prognosis

Greenstein AJ, Sachar DB, Gibas A, Schrag D, Heimann T, Janowitz HD, Aufses AH, Jr: Outcome of toxic dilation in ulcerative colitis and Crohn's colitis. *Gastroenterology* 1984; 86(part 2);1099.

Chapter 14. The Natural History of Inflammatory Bowel Disease

Meyers SJ, Janowitz, HD: The "natural history" of ulcerative colitis: An analysis of the placebo response. *J Clin Gastro* 1989;11:33–37.

Meyers S, Janowitz HD: The "natural history" of Crohn's disease: An analytic review of the placebo lesson. *Gastroenterology* 1984;87:1189–1192.

Chapter 15. Medical Management Programs

Janowitz HD: Therapeutic strategies in Crohn's disease. *Am J Gastro* 1978;70:667–690.

Meyers S, Janowitz HD: Crohn's disease, medical therapy. In Berk JE, Haubrich WS, Kalser MH, Roth JLA, Schaffner F (eds): *Bockus Gastroenterology*. Philadelphia, WB Saunders, 1985;2305–2317.

Janowitz HD, Bilotta J: Critical evaluation of medical therapy of inflammatory bowel disease. In Phillips S, Pemberton JH, Shorter RG (eds): *The Large Intestine, Physiology, Pathophysiology and Disease*. New York, Raven Press, 1991, pp 475–500.

Drug Therapy

Meyers S, Sachar DB, Goldberg JD, Janowitz HD: Corticotropin (ACTH) versus hydrocortisone in the intravenous treatment of ulcerative colitis. A prospective, randomized, double-blind clinical trial. *Gastroenterology* 1983;85:351–352.

Meyers S, Sachar DB, Present DH, Janowitz HD: Olsalazine in the treatment of ulcerative colitis patients intolerant of sulfasalazine. A prospective, randomized, placebo controlled, double blind, dose-ranging, clinical trial. *Gastroenterology* 1987;93:1255–1262.

Omega-3 Fatty Acids

Salomon P, Kornbluth A, Janowitz HD: Treatment of ulcerative colitis with fish oil omega-3 fatty acids: An open trial. *J Clin Gastro* 1990;12:157–161.

Chapter 18. How Effective Are Our Current Drugs

Kornbluth A, Salomon P, Sacks HS, Mitty R, Janowitz HD: Meta-analysis of the effectiveness of current drug therapy for ulcerative colitis. *J Clin Gastro* 1993;16:215–218.

Kornbluth A, Salomon P, Janowitz HD: The efficacy of current medical therapy in severe ulcerative colitis. An analytic review of the defined trials. (Abstract) *Am J Gastro* 1991;86:1356.

Salomon P, Kornbluth A, Aisenberg J, Janowitz HD: How effective are current drugs for Crohn's disease? A meta-analysis. *J Clin Gastro* 1992;14:211–215.

Chapter 21. The Results of Surgical Therapies

General

Janowitz HD: Problems in Crohn's disease: Evaluation of the results of surgical treatment. *J Chronic Dis* 1975;64:122–124.

Korelitz BI, Present DH, Alpert L, Marshak RH, Janowitz HD: Recurrent regional ileitis after ileostomy and colectomy for granulomatous colitis. *N Engl J Med* 1972;287:110–115.

Greenstein AJ, Sachar DB, Pasternack BD, Janowitz HD: Reoperation and recurrence in Crohn's colitis and ileocolitis: Crude and cumulative rates. *N Engl J Med* 1975;293:685–690.

Risk Factors

Wolfson DM, Sachar DB, Cohen A, Goldberg J, Styczynski R, Greenstein AJ, Gelernt IM, Janowitz HD: Granulomas do not affect postoperative recurrence rates in Crohn's disease. *Gastroenterology* 1982;83:405–409.

Sachar DB, Wolfson DM, Greenstein AJ, Goldberg J, Styczynski R, Janowitz HD: Risk factors for postoperative recurrence of Crohn's disease. *Gastroenterology* 1983;85:917–921.

Chapter 22. Postoperative Problems

Lerman B, Garlock JH, Janowitz HD: Suppurative pylephlebitis with multiple liver abscesses complicating regional ileitis: Review of literature—1940–1960. *Ann Surg* 1962;15:441–448.

Greenstein AJ, Sachar DB, Greenstein RJ, Janowitz HD, Aufses AH, Jr: Intraabdominal abscess in Crohn's (ileo) colitis. *Am J Surg* 1982;143:727–730.

Marshak RH, Korelitz BI, Klein SH, Wolf BS, Janowitz HD: Toxic dilation of the colon in the course of ulcerative colitis. *Gastroenterology* 1960;38:165–180.

Meyers S, Janowitz HD: The place of steroids in the therapy of toxic megacolon. *Gastroenterology* 1978;75:729–731.

Mogan G, Sachar DB, Bauer J, Janowitz HD: Toxic megacolon in ulcerative colitis complicated by pneumomediastinum. *Gastroenterology* 1980;559–562.

Greenstein AJ, Sachar DB, Gibas A, Schrag D, Heimann T, Janowitz HD, Aufses AH, Jr: Outcome of toxic dilation in ulcerative colitis and Crohn's colitis. *Gastroenterology* 1984; 86(part 2);1099.

Pouchitis

Subramani K, Harkaz N, Bilotta J, Bodian C, Rubin PH, Janowitz HD, Gelernt IM, Sachar DB: Refractory pouchitis: Does it reflect underlying Crohn's disease? *Gut* 1993;34:1539–1542.

Chapter 23. The Quality of Life and Inflammatory Bowel Disease

Meyers S, Walfish JS, Sachar DB, Greenstein AJ, Hill AG, Janowitz HD: Quality of life after surgery for Crohn's disease: A psychosocial assessment. *Gastroenterology* 1980;78:1–6.

Chapter 26. The Cancer Problem

Carcinoembryonic Antigen

Rule AH, Straus E, Vandervoorde J, Janowitz HD: Tumor-associated (CEA-reacting) antigen in patients with inflammatory bowel disease. *N Engl J Med* 1972;287:24–26.

Ulcerative Colitis

Greenstein AJ, Sachar DB, Smith H, Pucillo A, Vasiliades G, Kreel I, Geller SA, Janowitz HD, Aufses AH, Jr: Cancer in universal and left-sided ulcerative colitis: Clinical and pathological features. *Mt Sinai J Med* 1979;46:293–296.

Greenstein AJ, Sachar DB, Smith H, Pucillo A, Vasiliades G, Kreel I, Geller SA, Janowitz HD, Aufses AH, Jr: Cancer in universal and left-sided ulcerative colitis: Factors determining risk. *Gastroenterology* 1979;77:290–294.

Crohn's Disease

Brown N, Weinstein VA, Janowitz HD: Carcinoma of the ileum 25 years after bypass for regional enteritis: A case report. *Mt Sinai J Med* 1970;37(6);675–677.

Greenstein AJ, Janowitz HD: Cancer in Crohn's disease: The danger of bypassed loop. *Am J Gastro* 1975;64:122–124.

Greenstein AJ, Sachar DB, Pucillo A, Kreel I, Geller SA, Janowitz HD, Aufses AH, Jr: Cancer in Crohn's disease after diversionary surgery: A report of seven carcinomas occurring in excluded bowel. *Am J Surg* 1978;135:86–90.

Ulcerative Colitis and Crohn's Disease

Greenstein AJ, Sachar DB, Smith H, Janowitz HD, Aufses AH, Jr: Patterns of neoplasia in Crohn's disease and ulcerative colitis. *Cancer* 1980;46:403–407.

Greenstein AJ, Sachar DB, Smith H, Janowitz HD, Aufses AH, Jr: Comparison of cancer risk in Crohn's disease and ulcerative colitis. *Cancer* 1981;48:2742–2745.

Malignant Melanoma

Greenstein AJ, Sachar DB, Rosenberg IR, Lewis C, Ragu T, Szporn A, Janowitz HD: Malignant melanoma in inflammatory bowel disease. *Am J Gastro* 1992;87:317–320.

Lymphoma

Greenstein AJ, Mullin GE, Strauchen JA, Heiman T, Janowitz HD, Aufses AH, Sachar DB: Lymphoma in inflammatory bowel disease. *Cancer* 1991;69:1119–1123.

Chapter 27. Associated Diseases
Hemolytic Anemia

Altman A, Maltz C, Janowitz HD: Autoimmune hemolytic anemia in ulcerative colitis: Report of three cases, review of the literature and evaluation of modes of therapy. *Dig Dis Sci* 1979;24:282–285.

Myelogenous Leukemia

Fabry TL, Janowitz HD, Sachar DB: Acute myelogenous leukemia in ulcerative colitis. *J Clin Gastro* 1980;2:225–227.

Pancreatitis with Crohn's Disease

Meyers S, Gravenstein HJ, Cohen BA, Janowitz HD: Pancreatitis coincident with Crohn's ileocolitis. Report of a case and review of the literature. *Dis Col Rec* 1987;30:119–122.

Chapter 28. Extraintestinal Manifestations
General

Greenstein AJ, Janowitz HD, Sachar DB: Extraintestinal manifestations of Crohn's disease and ulcerative colitis: A study of 700 cases. *Medicine* 1976;55:401–412.

Meyers S, Janowitz HD: Crohn's disease, extraintestianl manifestations. In Berk JD, Haubrich WS, Kalser MH, Roth JLA, Schaffner F (eds): *Bockus Gastroenterology*. Philadelphia, WB Saunders, 1985, pp 2259–2267.

Mayer L, Janowitz HD: IBD (Crohn's disease and ulcerative colitis): Extraintestinal manifestations. In Hawkins AK, Williams A (eds): *Inflammatory Bowel Disease*. New York, Churchill Livingston, 1982, pp 121–130.

Mayer L, Janowitz HD: Extraintestinal manifestations. In Allan RN, Keighley MRB, Alexander-Williams J, Hawkins CF (eds): *Inflammatory Bowel Diseases*, ed. 2. Edinburgh, Churchill Livingston, 1990, pp 501–512.

Skin

Finkel SI, Janowitz HD: Trauma and the pyoderma gangrenosum of inflammatory bowel disease. *Gut* 1981;2:225–227.

Talansky A, Meyers S, Greenstein AJ, Janowitz HD: Does intestinal resection heal the pyoderma of inflammatory bowel disease? *J Clin Gastro* 1983;5:207–210.

Cryoglobulinemia

Altman A, Meyers S, Sachar DB, Janowitz HD: Crohn's ileocolitis, cutaneous gangrene and cryoglobulinemia. *Mt Sinai J Med* 1979;46:293–296.

Mayer L, Meyers S, Janowitz HD: Cryoproteinemia in the cutaneous gangrene of Crohn's disease: A report of two cases. *J Clin Gastro* 1981;3(1):17–21.

Renal Amyloidosis

Werther JL, Schapira A, Rubinstein O, Janowitz HD: Amyloidosis in regional enteritis: A report of five cases. *Am J Med* 1960;29:416–423.

Greenstein AJ, Sachar DB, Panday AKN, Dickman SH, Meyers S, Heiman T, Gumaste V, Werther LJ, Janowitz HD: Amyloidosis and inflammatory bowel disease, a 50-year experience with 25 patients. *Medicine* 1992;71:261–270.

Colchicine Therapy of Amyloidosis

Meyers S, Janowitz HD, Gumaste V, et al: Colchicine therapy of the renal amyloidosis of ulcerative colitis. *Gastroenterology* 1988;94:1503–1507.

Obstructive Hydronephrosis

Present DH, Rabinowitz JG, Banks PA, Janowitz HD: Obstructive hydronephrosis: A frequent but seldom recognized complication of granulomatous disease of the bowel. *N Engl J Med* 1969;280:523–528.

Metastatic Crohn's Disease

Lebwohl M, Fleischmajer R, Janowitz HD, Present D, Priolequ PG: Metastatic Crohn's disease. *J Am Acd Derm* 1984;1(10):33–38.

A Selected Annotated Bibliography

Chapter 1. Introduction

Kirsner JB, Shorter RG (eds): *Inflammatory Bowel Disease*, ed. 2. Philadelphia, Lea & Febiger, 1988.
Allan RN, Keighley MRB, Alexander-Williams J, Hawkins CF (eds): *Inflammatory Bowel Diseases*, ed. 2. Edinburgh, Churchill Livingston, 1990.
An excellent comprehensive monograph edited from Birmingham with a strong British and European flavor paralleling the pioneer important American text of Kirsner and Shorter.

Historical

Wilks S, Moxon W: *Lectures on Pathological Anatomy*, ed. 2. London, Churchill, 1875.
Generally, and probably correctly, given credit as being the first coherent description of ulcerative colitis.
Dalziel AM: Chronic interstitial enteritis. *Br Med J* 1913;2:1068–1070.

This early clear-cut description of what was to be called Crohn's disease, although published in a leading journal, failed to elicit the appropriate recognition, perhaps because so few new cases were appearing at that time.

Crohn BB, Ginzburg L, Oppenheimer GD: Regional ileitis. *JAMA* 1932;99:1323–1329.

The classical description. Surprisingly few clinical features have been added except for the colonic involvement. The absence of gross and microscopic photographs is puzzling.

Ginzburg L, Oppenheimer GD: Non-specific granulomata of the intestines. *Ann Surg* 1933;27:1046–1062.

A neglected paper, paralleling their publication with Crohn, rich in pathologic evidence and noting colonic involvement.

Chapter 2. Disease or Diseases?

Lockhart-Mummery HE, Morson BC: Crohn's disease (regional enteritis) of the large intestine and its distinction from ulcerative colitis. *Gut* 1960;1:87–105.

Although others, including Marshak, Wells, and Brooke among them, had been pointing to colonic involvement in regional enteritis, this was the seminal modern paper clearly separating Crohn's colitis from ulcerative colitis.

Jones TH, Lennard-Jones JE, Morson BC, Chapman M, Sacklin MJ, Sneath PHA, Spicer CC, Card WI: Numerical taxonomy and discriminant analysis applied to non-specific colitis. *Q J Med* 1973;42:715–732.

A pioneer attempt to separate ulcerative colitis from Crohn's disease by the technique of numerical taxonomy, worth rereading in the light of the current interest in the "indefinite" groups of nonspecific IBD. Discriminant analysis reduced to five the criteria that most sharply separated ulcerative colitis from Crohn's disease.

Hodgson HJF: One disease or two? In Allan RN, Keighley MRB, Alexander-Williams J, Hawkins CF (eds): *Inflammatory Bowel Diseases*, ed. 2. Edinburgh, Churchill Livingston, 1990, pp 121–126.

An interesting effort to bring in the Scotch verdict of "Not Proven" in regard to the identity or difference of the two basic disorders.

Chapter 3. The Nature of the Beasts

Hadfield G: The primary histological lesion of regional ilietis. *The Lancet* 1939;2:773–775.

The pioneer histologic description of regional enteritis complementing the original clinical description with emphasis on the granulomatous (sarcoid) nodule.

Vascular Factor in Crohn's Disease

Wakefield AJ, Sawyer AM, Dhillon AP, Patillo RM, Rawler PM, Lewis AAM, et al: Pathogenesis of Crohn's disease: Multifocal gastrointestinal infarction. *The Lancet* 1989;1857–1862.

The first of a series of papers from these writers focusing on the role of vascular lesions to account for the early pathologic lesions of this disorder.

Sankey EA, Dhillon AP, Anthony A, Wakefield AJ, Sim R, More L, Hudson M, Sawyer AM, Proinder RE: Early mucosal changes in Crohn's disease. *Gut* 1993;34:375–381.

The authors conclude that damage to small mucosal capillaries and their rupture occurs before invasion of the lamina propria by inflammatory cells and loss of the overlying epithelium seems to follow this vascular damage and produces the early aphthoid lesion of Crohn's disease.

Chapter 4. Theories of Etiology and Their Therapeutic Implications

General

Sachar DB, Walfish J, Auslander MD: Etiological theories of inflammatory bowel disease. *Clin Gastro* 1980;9:231–258.

A comprehensive catalogue of research and speculation over the entire spectrum of nonspecific IBD of the prior decade.

Immunology

MacDermott RP, Stenson WF: Alternations of the immune system in ulcerative colitis and Crohn's disease. *Immunol* 1988;42: 285–328.

A sophisticated account of the complex machinery involved in the alterations of the immune system in IBD.

Mayer L, Eisenhardt D, Salomon P: Expression of class II molecules on intestinal epithelial cells in humans. *Gastroenterology,* 1991;100:3–12.

Mayer L, Eisenharde D: Lack of induction of suppressor T cells by intestinal epithelial cells from patients with inflammatory bowel disease. *J Clin Invest* 1990;86:1255.

Two pioneering studies in the function and dysfunction of the intestinal epithelial cells in the immune reactions of the gut in IBD.

Genetics

Rotter JI, Yang H: Resolving the genetics of IBD: The challenge for the 90s. *Progress in Inflammatory Bowel Disease* 1993; 14:1–7.

A brilliant analysis of the newer information on the genetics of IBD with emphasis on current research in identifying the susceptibility genes.

Family Studies

Bennett RA, Rubin PH, Present DH: Frequency of inflammatory bowel disease in offspring of couples presenting with inflammatory bowel disease. *Gastroenterology* 1991;100:1638–1643.

The frequency of IBD in children was slightly higher if both parents had already developed IBD at the time of conception (67%) contrasted to 50% when only one parent or neither had developed IBD when conception occurred.

Twin Studies

Tysk C, Linberg E, Jänerot G, Flodérus-Myrhed B: Ulcerative colitis and Crohn's disease in an unselected population of monozygotic and dizygotic twins. A study of heritability and the influence of smoking. *Gut* 1988;29:990–996.

Heredity as an etiological factor was stronger in Crohn's disease than in ulcerative colitis. Monozygotic twins with Crohn's disease were more likely to be smokers than monozygotic twins with ulcerative colitis.

Serum Markers of IBD

Shanahan R, Duerr RH, Rotter JI, Yang I, Sutherland LR, McElree C, Landers CJ, Targan SR: Neutrophil autoantibodies in ulcerative colitis, familial aggregation and heterogeneity. *Gastroenterology* 1992;103:456–461.

An important report on the significance of this serum marker of ulcerative colitis and its family distribution.

Transmission Studies

Mitchell DN, Rees RJW: Agent transmissible from Crohn's disease tissue. *The Lancet* 1970;2:168–171.

Although this pioneer report on transmission of an agent from Crohn's disease tissue did not lead to a clear-cut solution of Crohn's etiology, it served as a most important stimulus for the hot pursuit of an infectious agent which still persists.

Microbiology

Gorbach SL: Speculation on the role of microorganisms in the etiology of IBD in internal microflora in IBD: Implications for etiology and therapy. In Kirsner JB, Shorter RG (eds): *Inflammatory Bowel Disease*. Philadelphia, Lea & Febiger, 1980, pp 66–67.

Mycobacteria and Crohn's Disease

Stainsby KJ, Lowes JR, Allan RN, Ibbotson JP: Antibodies to *Mycobacterium paratuberculosis* and nine species of environmental mycobacterium in Crohn's disease and control subjects. *Gut* 1993;34:371–374.

Further evidence that the presence of mycobacterium in the tissues of patients with Crohn's disease may be the result of secondary invasion of a previously damaged mucosa.

C. difficile Toin

Trnka YU, LaMont JJ: Association of *Clostridium difficile* toxin with symptomatic relapse in chronic inflammatory bowel disease. *Gastroenterology* 1981;80:692–696.
 The case for incriminating *C. difficile* toxin as involved in or actuating clinical relapse in IBD is to be contrasted with the paper of Meyers et al. (*Gastroenterology* 1981;80:697–702), and should be read in the light of John Bartlet's judicious editorial (*Gastroenterology* 1981;80:863–865). I believe the case has not been made except in the rarest situation.

Role of the Fecal Stream

Harper PH, Lee ECG, Kettlewell MGW, Bandt MK, Jewell DP: Role of the fecal stream in the maintenance of Crohn's colitis. *Gut* 1985;26:279–284.

Rutgeerts P, Geboes K, Peters M, Hiele M, Pennick F, Aerts R, Kerremans R, Vantrappen G: Effect of fecal stream diversion on recurrence of Crohn's disease in the neoterminal ileum. *The Lancet*, 1991;338:771–774.
 Ever since the early publications from Mount Sinai Hospital on the effects of diverting the fecal stream by exclusion operations (Ginsburg L, Colp R, Sussman L: Ileostomy without exclusion. *Ann Surg* 1939;110:448), clinical observers including myself have postulated that among the factors involved in Crohn's disease is the movement downstream of some injurious substance or substances.
 The two papers cited here are ingenious attempts to demonstrate this phenomenon by experiments in man. The first concluded that factors greater than 0.22 microns in the fecal stream are responsible for the maintenance and exacerbation of inflammation in Crohn's disease. The removal by surgical manipulation or diverting the fecal stream away from newly formed neo-ileum and anastomotic ileocolostomy protected that area temporarily from recurrences for at least six months, according to the second paper.

Psychological

Engel GL: Studies in ulcerative colitis III: The nature of the psychologic processes. *Am J Med* 1955;19:231–256.

George Engel, in this one of a series of studies on ulcerative colitis, brings his subtle and intuitive psychoanalytically trained mind to the care of patients from the viewpoint of a well-trained internist; the outstanding writer from this point of view.

Monk M, Mendeloff AI, Siegel CI, Lilenfeld A: An epidemiological study of ulcerative colitis and regional enteritis among adults in Baltimore. III Psychological and possible stress-precipitating factors. *J Chron Dis* 1970;22:565–578.

Helzer JE, Stillings WA, Chammas S, Norland CC, Alpers DH: A controlled study of the association between ulcerative colitis and psychologic diagnoses. *Dig Dis Sci* 1982;27:1513–1518.

Helzer JE, Chammas S, Norland CC, Stillings WA, Alpers DH: A study of the association between Crohn's disease and psychiatric illness. *Gastroenterology* 1984;86:324–330.

Among the welter of impressionistic papers regarding the role of psychologic factors in IBD, the studies by Monk and colleagues and Alpers and his co-workers are salutary attempts to bring new rigorous approaches, including statistical methods and better defined psychologic terms to the problem.

Interestingly, Alpers's studies revealed no relationship between psychiatric diagnoses and ulcerative colitis, but did find one with Crohn's disease, especially depression.

The etiologic role of psychic factors seems to have fallen by the wayside in recent years.

Chapter 5. Modes of Clinical Presentation

Farmer RG, Hawk WA, Turnbull CD: Clinical patterns in Crohn's disease: A statistical study of 615 cases. *Gastroenterology* 1975;68:627–635.

An important attempt to sort out the variety of clinical patterns with which Crohn's disease presents itself in a large number of patients and relate the subsequent course and prognosis to the initial anatomical involvement.

Natural History

Farmer RG, Easley KA, Rankan GB: Clinical patterns, natural history, and progression of ulcerative colitis: A long-term follow-up of 1116 patients. *Dig Dis Sci* 1993;38:1137–1146.

An important study of the life history of ulcerative colitis, notable for its long follow-up. The data on progression of the disease are extremely valuable.

Crohn's Disease of the Appendix

Ruiz V, Unger SW, Morgan J, Wallace MK: Crohn's disease of the appendix. *Surgery* 1990;107:113–117.

"Metastatic" Crohn's Disease

Strum DT, Guenther L: Metastatic Crohn's disease. Case report and review of the literature. *Arch Dermatol* 1990;126:645–648.

Ulcerative Colitis and Crohn's Disease in Childhood

Booth IW, Harries JT: Imflammatory bowel disease in childhood. *Gut* 1984;25:188–202.
A concise review of IBD from the British pediatric gastroenterologist's point of view. The sections on nutritional support and growth failure in Crohn's disease are a nice summary of current information.

Chapter 7. The Essential Investigations

Radiographic

Marshak RH, Lindner AE: Ulcerative colitis and granulomatous colitis. *J Mt Sinai Hosp* 1968;33:444–502.
Marshak RH, Lindner AE: The radiologic diagnosis of chronic ulcerative colitis and Crohn's disease. In Kirsner JB, Shorter RG (eds): *Imflammatory Bowel Disease*. Philadelphia, Lea & Febiger, 1980, pp 341–409.
These articles of monographic proportions are a monument of a lifetime's radiographic study of IBD. Newer imaging techniques including endoscopic ones will add to their storehouse of information but will not replace it.
Jones TH, Lennard-Jones JE, Young AC: Reversibility of radiological appearance during clinical improvement of colonic Crohn's disease. *Gut* 1969;10:738–743.
Ulcerative colitis has been long reported as demonstrating radio-

logical reversibility (Goldberg HT, Carbone JV, Margulis RR: Roentgenographic reversibility in ulcerative colitis in children treated with steriod enemas. *AJR* 1968;103:365–379). This paper was a pioneer in presenting evidence of reversibility in Crohn's disease, now a commonplace.

Endoscopy

Waye, JD: Endoscopy in inflammatory bowel disease. *J Clin Gastro* 1980;9:279–296.

Laufer I, Mullens JE, Hamilton J: Correlation of endoscopy and double contrast radiography in the early stages of ulcerative and granulomatous colitis. *Radiology* 1976;118:1–5.

Rectal Biopsy

Goodman MJ, Kirsner JB, Ridell RH: Usefulness of rectal biopsy in inflammatory bowel disease. *Gastroenterology* 1977;72: 952–956.

Rotterdam H, Korelitz BI, Sommers SC: Micro granuloma in grossly normal rectal mucosa in Crohn's disease. *Am J Clin Pathol* 1977;64:550–554.

Surawicz CM, Meisel JL, Ylvisaker T, Saunders DR, Rubin CE: Rectal biopsy in the diagnosis of Crohn's disease: Value of multiple biopsies and serial sectioning. *Gastroenterology* 1981;80:66–71.

Further studies demonstrating convincingly the presence of microscopic Crohn's disease in grossly normal rectums, consistent with our earlier study (*Gastroenterology* 1967:52, 1113).

Chapter 10. Differential Diagnosis and Errors in Diagnosis

Persson S, Danielson D, Kjellander J, Wallenstein S: Studies of Crohn's disease I. The relationship between *Yersinia enterocolitica* infection and terminal ileitis. *Acta Chir Scand* 1976;142:84–90.

A long-term study (five to eight years) that demonstrated that terminal ileitis associated with *Yersinia* infection did not develop into Crohn's disease of the ileum.

Diversion Colitis

Roe AM, Warren BF, Brodribb AJM, Brown C: Diversion colitis and involution of the defunctionalized anorectum. *Gut* 1993;34:382–385.

A nice study which documents changes in the histology of the defunctionalized anorectum following operation for non-IBD; these points are relevant to the diagnoses of diversion colitis.

Collagenous and Lymphocytic Colitis

Lindstrom CG: "Collagenous colitis" with watery diarrhea—a new entity? *Pathol Eur* 1976;11:87–89.

The pioneer paper by the Swedish pathologist who reported the first case, and invented the term *collagenous*. More than 300 cases have been subsequently reported in the literature.

Giardiello FM: A review of atypical colitides: Collagenous and lymphocytic colitis. *Progress in Inflammatory Bowel Disease* 1993;14:1–4.

A succinct review of current opinion regarding these possibly related entities.

Chapter 12. Assessing the Patient's Clinical Status

Lennard-Jones JE, Ritchie JK, Hilder W, Spicer CC: Assessment of severity in colitis: A preliminary study. *Gut* 1975;16:579–584.

A strong case for a simple classification of severity in acute ulcerative colitis with predictive value for four features: temperature, pulse rate, bowel frequency, and plasma albumen. This noteworth attempt now needs prospective application.

Best WR, Bechtel JM, Singleton JW: Rederived values of the eight coefficients of the Crohn's disease activity index (CDAI). *Gastroenterology* 1979;77:843–846.

Despite its limitations the Crohn's Disease Activity Index (CDAI) remains a landmark in the effort to quantitate the disease activity for clinical research purposes, and has stimulated most current attempts to find simpler clinical indices for daily medical use.

Chapter 13. Prognosis

Ulcerative Colitis

Edward FC, Truelove SC: The course and prognosis of ulcerative colitis: Part I, Short-term prognosis. *Gut* 1963;4:299–308. Part II, Long-term prognosis. *Gut* 1963;4:309–315.

Part I: Unique 100% follow up of 624 patients with ulcerative colitis between 1938 and 1962, emphasizing the chief factor affecting the fatality rate in the first referred attack of the disease.

Part II: Directed to long-term prognosis. An outstanding careful follow up in ulcerative colitis, emphasizing the high mortality in the first severe attack and likelihood of an improved long-term course if the patient survives.

Proctitis

Powell-Tuck J, Ritchie JK, Lennard-Jones JE.: The prognosis of ideopathic proctitis. *Scand J Gastro* 1977;12:727–732.

A long-term study of 189 patients seen over a 10-year period. Excellent data that shows that survival is little affected by the disease. Extension above the iliac crest was 5% at five years and 12% at 10 years. The possibility of radical surgical treatment was 3 to 5%.

Toic Dilation

Jalan KN, Sircus W, Card WI, Falconer CWA, Bruce J, Creary GP, McManus JPA, Small WP, Smith AN: An experience of ulcerative colitis and toxic dilation in 55 cases. *Gastroenterology* 1969;57:68–82.

Katzka I, Katz S, Morris E: Management of toxic megacolon: The significance of early recognition in medical management. *J Clin Gastro* 1979;1:307–311.

Fazio VF: Toxic megacolon in ulcerative colitis and Crohn's disease. *J Clin Gastro* 1980;9:389–407.

An interesting group of contrasting experiences in the management of toxic dilation of ulcerative colitis, earlier on and more recently. Early recognition and, more important, preventing this complication are critical, but it remains to be seen whether aggressive medical management of subsequent colitis reduces the need for urgent colectomy.

Crohn's Disease

Rutgeerts P, Geboes K, Vantrappen G, Beyls J, Kerremans R, Hiele M: Predictability of the postoperative course of Crohn's disease. *Gastroenterology* 1990;99:956–963.

One of the studies that emphasizes the early recurrence of endoscopic and microscopic evidence of disease which is not correlated with symptomatic activity. The group with aggressive Crohn's disease recurrence is similar to those reported from Mount Sinai by Sachar et al. (*Gut* 1988;29:588–592).

Predicting Outcome in Children: Crohn's Disease

Gryboski TD, Spiro HM: Prognosis in children with Crohn's disease. *Gastroenterology* 1978;74:807–817.

Puntis J, McNeish AS, Allan RN: Long-term prognosis of Crohn's disease with onset in childhood and adolescence. *Gut* 1984;25:329–336.

Two splendid reviews of large series of patients carefully followed for long periods of time. The experience in New Haven should be compared to the findings in Birmingham. At Yale most patients received steroids and fewer were operated on than in Birmingham, where steroids were used most sparingly and the surgical attack was more aggressive. The latter reflects the influence of Trevoir Cooke, whose attitudes are presented forcefully in "Factors in the management of Crohn's disease: A discussion paper." *J Roy Soc Med* 1981;74:753–758.

Chapter 15. Medical Management Programs

General

Peppercorn MA: Advances in drug therapy for inflammatory bowel disease. *Ann Int Med* 1990;112:50–60.

A thoughtful judicious summary of the current available drugs for the treatment and maintenance of remission in IBD and finds a place for ACTH in the treatment of acute severe ulcerative colitis.

Ulcerative Colitis

CORTISONE

Truelove SC, Witts, LJ: Cortisone in ulcerative colitis: Final report on a therapeutic trial. *Br Med J* 1955;2:1041–1048.

The final report of the pioneer controlled trial of cortisone in ulcerative colitis, which gave the imprimatur to the use of this drug and justified its usefulness. The dosages used (up to 100 mg/day/for several weeks) were probably much too low by current standards.

ACTH VS. CORTICOTROPHIN

Truelove SC, Witts, LJ: Cortisone and corticotrophin in ulcerative colitis. *Br Med J* 1959;1:387–394.

In this early study, corticotrophin (80 units daily) proved more effective than cortisone (50 mgs daily) in bringing about a complete remission in the course of six weeks, yet the authors concluded that there was "little to choose between corticotrophin and cortisone in the treatment of a first attack of ulcerative colitis."

Kaplan HD, Portnoy B, Binder HT, Amatruda J, Spiro H: A controlled evaluation of intravenous adrenocorticoatrophic hormone and hydrocortisone in the treatment of acute colitis. *Gastroenterology* 1975;69:91–95.

This study of 22 patients with a mixture of ulcerative colitis and Crohn's disease with acute colitis showed no difference in response to ACTH or hydrocortisone. It should be compared to the study of Meyers et al. (*Gastroenterology* 1983;85:351–357), which demonstrated clear-cut differences in 66 patients with only ulcerative colitis in the response to these two drugs depending on their previous exposure to one or the other.

TOPICAL STEROIDS

Mulcher CJJ, Tytgat CNJ: Review article: Topical corticosteroids in inflammatory bowel disease. *Aliment Pharmacol Ther* 1993; 17:125–130.

A succinct review of the status of topical steroid drugs in inflammatory bowel disease with emphasis on the current development of those with minimal systemic corticosteroid activity.

IMMUNOSUPPRESSANTS

Twell DP, Truelove SC: Azathioprine in ulcerative colitis: Final report on a controlled therapeutic trial. *Br Med J* 1974; 4:627–630.

In this study on the treatment of a first acute episode of ulcerative colitis, the addition of azathioprine to a standard corticoid program confirmed no benefit, nor when continued on a maintenance program in these patients. It did confirm benefit in patients maintained on the drug following a relapse of established disease.

I concur completely with the authors' suggestion that azathioprine, if it has any place in ulcerative colitis, is as a maintenance drug in patients "who have not done well on corticosteroid or sulfasalazine and in whom there is good reason for not performing or deferring proctocolectomy."

SALAZOPYRINE

Svartz N: Salazopyrine, a new sulfanilamide:
  A. Therapeutic results in rheumatic polyarthritis.
  B. Therapeutic results in ulcerative colitis.
  C. Toxic manifestations in treatment with sulfanilamide preparation. *Acta Med Scand* 1942;110:557–590.
Svartz N: The treatment of 124 cases of ulcerative colitis with salazopyrine and attempts at desensitization in cases of hypersensitization to sulfa. *Acta Med Scand* 1948;206 (*suppl*):465–472.

The first reports on the use of salazopyrine in ulcerative colitis by its introducer. Very well worth rereading especially the comments on the need for more prolonged maintenance therapy.

MAINTENANCE THERAPY

Dissanayake AS, Truelove SC: A controlled therapeutic trial of long-term maintenance treatment of ulcerative colitis with sulfasalazine (salazopyrine). *Gut* 1973;14:923–926.

The forceful argument for maintaining patients with ulcerative colitis in remission on salazopyrine indefinitely.

Azad Khan AK, Howes DT, Piris J, Truelove SC: An optimum dose of sulfasalazine for maintenance treatment in ulcerative colitis. *Gut* 1980;21:232–240.

Evidence regarding the long-term need of this drug for maintenance therapy limited only by the frequency of adverse drug reactions.

MODE OF ACTION: SULFASALAZINE

Azad Khan AK, Piris J, Truelove SC: An experiment to determine the active therapeutic moiety of sulfasalazine. *The Lancet* 1977;2:892–895.

Truelove's ingenious study which demonstrated that 5-ASA was the active moiety of sulfasalazine, opening up the flood of newer 5-ASA forms.

SULFASALAZINE, PREDNISONE, TOPICAL CORTICOSTEROID

Lennard-Jones JE, Langmore AJ, Newell AC, Wilson CWE: An assessment of prednisone, salazopyrine and topical hydrocortisone hemisuccinate used as outpatient treatment for ulcerative colitis. *Gut* 1960;1:217–222.

Salazopyrine approached the effectiveness of prednisone in this outpatient treatment trial; not surprisingly the patients with ulcerative colitis involving all or part of the colon distal to the splenic flexure did poorly on hydrocortisone.

SULFASALAZINE REVIEWED

Sutherland LR, May GR, Shaffer EA: Sulfasalazine revisited: A meta-analysis of 5-aminosalicylic acid in the treatment of ulcerative colitis. *Ann Int Med* 1993;118:540–549.

A beautifully organized meta-analysis which reveals, not surprisingly, that, except for the sulfasalazine-sensitive patient, the newer forms of delivering 5-ASA to the colon confer no therapeutic advantage over the classic preparation of Azulfidine® for the treatment of active disease or the maintenance of remission.

CYCLOSPORINE

Lichtiger S, Present DH: Preliminary report: Cyclosporine in the treatment of severe active ulcerative colitis. *The Lancet* 1990;2:16–19.

Convincing evidence of the value of cyclosporine in the treatment of sick patients who have failed IV steroid therapy for severe acute ulcerative colitis, which is substantiated by the more recent controlled trial.

FISH OIL IN ULCERATIVE COLITIS

Stenson WT, Cort D, Rodgers J, et al: Dietary supplementation with fish oil in ulcerative colitis. *Ann Int Med* 1993;116:609–614.

Hawthorne AB, Daneshmene TK, Hawkey CJ: Treatment of ulcerative colitis with fish oil supplementation: A prospective 12-month randomized controlled trial. *Gut* 1992;33:992–998.

Both studies report only minimal efficacy with the use of fish oil in the form of maxiEPA, while super evening primrose oil may confer some benefit (*Aliment Pharmacol Ther* 1993;7:159–166).

Crohn's Disease

SALAZOPYRINE, PREDNISONE, AZATHIOPRINE

Summers RW, Switz DM, Session JT Jr, et al: National Cooperative Crohn's Disease Study: Results of drug treatment. *Gastroenterology* 1979;77:846–869.

The classic controlled trial of treatment of Crohn's with salazopyrine, prednisone, and azathioprine. As a member of the Advisory Committee, I agree that the four-month trial was rather short and that the decision to stop azathioprine (because of five cases of pancreatitis) was premature. The maintenance study cannot be faulted.

Melchow H, Ewe K, Brandes JW, et al: European Cooperative Crohn's Disease Study (ECCDS): Results of drug treatment. *Gastroenterology* 1984;86:249–266.

Essentially confirmation of the U.S. study.

COMBINATION THERAPY: SALAZOPYRINE AND STEROIDS

Singleton JW, Summers RW, Kern F, Becklel TM, Bert WR, Hansen RN, Winship DH: A trial of sulfasalazine as adjunctive therapy in Crohn's disease. *Gastroenterology* 1979;77:887–897.

IMMUNOSUPPRESSANT THERAPY

Brooke ZW, Case DR, King DW: Place of azathioprine for Crohn's disease. *The Lancet* 1976;1:1041–1042.
This pioneer study gave the impetus to treat Crohn's disease with immunosuppressant drugs.

Present DH, Korelitz BI, Wisch N, Glass JL, Sachar D, Pasternack BS: Treatment of Crohn's disease with 6-mercaptopurine: A long-term, randomized double-blind study. *N Engl J Med* 1980;302:981–987.
This paper was the principal impetus to the modern use of 6-MP in Crohn's disease, new widely used and accepted.

Kozarek RA: Review article: Immunosuppression study for inflammatory bowel disease. *Aliment Pharmacol Ther* 1993;7:117–123.
A succinct review of the current status of immunosuppressive drugs in both Crohn's and ulcerative colitis with emphasis on azathioprine and 6-mercaptopurine.

5-ASA PROPHYLAXIS IN CROHN'S DISEASE POSTOPERATIVELY

Wenchert A, Kristensen M, Eklund M, et al: The long-term prophylactic effect of sulfasalazine (salazopyrine) in primary resected patients with Crohn's disease. A controlled double-blind trial. *Scand J Gastro* 1978;13:161–167.
One of the earliest papers on the prophylactic value of sulfasalazine in resected patients with Crohn's.

Ewe K, Herforth C, Malchow H, Tedinsky HJ: Postoperative recurrence of Crohn's disease in relation to radicality of operation and sulfasalazine prophylaxis: A multicentric trial. *Digestion* 1989;42:224–232.
A more recent paper with evidence on the value of sulfasalazine in preventing clinical recurrence postoperatively.

International Mesalazine Study Group. Coated oral 5-aminosalicylic acid versus placebos in maintaining remission of inactive Crohn's disease. *Aliment Pharmacol Ther* 1990;4:55–64.
Further convincing evidence of the value of this approach at least for the first year following operation. The need for a fresh

reappraisal of this subject is well presented in the Editorial: 5-aminosalicylates for prevention of recurrence in patients with Crohn's disease: Time for reappraisal by L. R. Sutherland. *J Clin Gastro* 1991;13:5–7.

MAINTENANCE IMMUNOSUPPRESSANT DRUGS

O'Donoghue DP, Dawson AM, Powell-Tuck J, Brown RL, Lennard-Jones JE: Double-blind withdrawal trial of azathioprine in maintenance treatment for Crohn's disease. *The Lancet* 1978;2:955–957.

Maintenance therapy in Crohn's disease with azathioprine was effective during the period of remission. The first clear-cut demonstration of this effect.

6 MP-TOXICITY

Present DH, Meltzer SJ, Wolke A, Korelitz BI: short and long-term toxicity to 6-mercaptopurine in the treatment of inflammatory bowel disease. *Gastroenterology* 1985;88:1545.

A reassuring report on the safety of this immunosuppressant drug.

METRONIDAZOLE

Ursing B, Alm T, Barany F, Berglen I, Ganrot-Nortin K, Hoeveln J, et al: A comparative study of metronidazole and sulfasalazine for active Crohn's disease. The Cooperative Crohn's Disease Study in Sweden (CCDSS). *Gastroenterology* 1982; 83:550–562.

A good controlled trial on Crohn's disease in 78 patients in a crossover study demonstrating that metronidazole was slightly more effective than salazopyrine in a five-month period. It seems worthwhile switching to metronidazole from sulfasalazine when it fails, but not the reverse.

METHOTREXATE IN ULCERATIVE COLITIS
AND CROHN'S DISEASE

Kozarek RA: Review article: Immunosuppressive therapy for inflammatory bowel disease. *Aliment Pharmacol Ther* 1993;7: 117–123.

An excellent account of the status of methotrexate as therapy in both ulcerative colitis and Crohn's disease by its original introducer.

Chapters 19–22. Surgical Management

The Decision for Surgery

Farmer RG, Haworth WA, Turnbull RB Jr: Indications for surgery in Crohn's disease: Analysis of 500 cases. *Gastroenterology* 1976;71:245–250.

In this study, the indication for surgical intervention in Crohn's disease was related to the anatomic disease locations (and clinical patterns), and special emphasis was placed on the varied course of the disease—a well-needed emphasis.

The Choice of Operations
Ulcerative Colitis

ILEOSTOMY

Brooke BN: The management of an ileostomy including its complications. *The Lancet* 1952;2:102–104.

The master on the Brooke ileostomy.

CONTINENT ILEOSTOMY OF KOCK

Kock, NGL: Intraabdominal "reservoir" in patients with permanent ileostomy. *Arch Surg* 1969;99:223–231.

Gelernt IM, Bauer JT, Kreel I: Reservoir ileostomy: Early experience with 54 patients. *Ann Surg* 1977;185:179–184.

Kelly KA, Phillips SF, Beahrs OH: The continent ileostomy. In Kirsner JB, Shorter RG (eds): *Inflammatory Bowel Disease*. Philadelphia, Lea & Febiger, 1980, pp 622–642.

The original description and two American experiences with Kock's continent ileostomy, which has little place, in my opinion, in the surgery for ulcerative colitis.

ILEOANAL ANASTOMOSIS

Martin LW, LeCoutre C, Schubert WK: Total colectomy and mucosal proctectomy with preservation of continence in ulcerative colitis. *Ann Surg* 1977;186:477–479.

Parks AG, Nicholls RJ, Belliveau P: Proctocolectomy with ileal resection and anal anastomosis. *Br J Surg* 1980;67:533–538.

Utsunomiya T, Iwarma M, et al: Total colectomy, mucosal proctectomy and ileoanal anastomosis. *Dis Col Rec* 1980;23:459–466.

The continuing evolution of the operation designed to preserve the rectal and anal musculature with a pelvic reservoir.

Kelly KA, Pemberton IH, Wolff BG, Dozois RR: Ileal pouch—anal anastomosis. *Current Problems in Surgery* 1992;29(2):52–131.

A magnificent and highly readable short monograph on the 1200 patients treated at the Mayo Clinic with the ileal J-pouch—anal anastomosis for ulcerative colitis, with its candid analysis of success and failures, including the problems of "pouchitis," and includes a preliminary report on the alternative operation: the ileal pouch—distal rectal anatomosis.

Fujita S, Jusunoki M, Shoji Y, Owada T, Utsunomiya J: Quality of life after total proctocolectomy and ileal J-pouch anal anastomosis. *Dis Col Rec* 1992;35:1030–1039.

An attempt to add the factors of lifestyle and predominant personality traits into the analysis of the satisfactoriness of this now leading operation for ulcerative colitis.

POUCHITIS

Svaninger G, Nordgren S, Oresland T, Hulton L: Incidence and characteristics of pouchitis in the Kock continent ileostomy and the pelvic pouch. *Scan J Gastroenterol* 1993;28:645–700.

A careful prospective study of this important complication of the pouch, which sounds a note of caution regarding the possibility of an increasing incidence of pouchitis in the future.

Crohn's Disease

MINIMAL SURGERY AND STRICTUROPLASTY FOR CROHN'S DISEASE

Lee ECG, Papaioannou N: Minimal surgery for chronic obstruction in patients with extensive or universal Crohn's disease. *Ann R Coll Surg Engl* 1982;64:229–233.

Alexander-Williams J, Fornaro M: Stricturoplasty beim morbus Crohn's. *Der Chirug* 1982;53:799–801.

Pleas for increasing more restrictive surgery in Crohn's disease with emphasis on stricturoplasty rather than resection in stenotic small bowel Crohn's disease, now well accepted.

Results of Surgery in Crohn's Disease

Lennard-Jones JE, Stalden GA: Prognosis after resection for chronic regional enteritis. *Gut* 1967;8:332–336.

This was an important contribution since it defined the types of recurrences after resection, and even more important, it applied for the first time the method of life tables for the problem of results of surgery. No paper devoted to follow up results of any surgical therapy since then can be seriously considered unless it uses the actuarial method.

Dombal FT, Burton I, Goligher JC: Recurrence of Crohn's disease after primary excisional surgery. *Gut* 1971;12:519–527.

A pioneer study on recurrence rates after resection in Crohn's disease using actuarial methods that pointed out the high five-year recurrence rate (45% for ileal disease, and a bit lower, 35%, for colonic disease).

Hellers C: Crohn's disease in Stockholm County (1950–1974). A study of epidemiology. Results of surgery and long-term prognosis. *Acta Chir Scand* 1979;490(suppl):1–84.

A wonderfully well-organized study with a well-defined and collected catchment base.

Longo WE, Oakley JR, Lavery IC, Church JM, Fazio V: Outcome of ileorectal anastomosis for Crohn's colitis. *Dis Col Rec* 1992;35:1066–1071.

This report reinforces the generally held dictum that patients with Crohn's disease should have an ileorectal anastomosis as an alternative to proctocolectomy if the rectum is not seriously diseased and sphincteric function not compromised.

Lock MR, Farmer RG, Fazio VW, Tagelman DC, Lavey IC, Weakly FL: Recurrence and reoperation for Crohn's disease. The role of disease location in prognosis. *N Engl J Med* 1981; 304:1580–1588.

Further emphasis on the high recurrence rate in ileocolitis and ileitis, and the lower rate in colitis; the role of anastomosis in recurrence rate in Crohn's colitis is not touched on, however.

Ambrose NS, Keighley MRB, Alexander-Williams J, Allan RN: Clinical impact of colectomy and ileorectal anastomosis in the management of Crohn's disease. *Gut* 1984;25:223–227.

A nice study reporting both cumulative recurrence and cumulative reoperation rates (life table method). The recurrence rate of 64%, with reoperation rate of 48% at 11 years is in keeping with our own data (*N Engl J Med* 1975;293:685–690) for this type of operation. Like our report (*Gastroenterology* 1983;85:917–21) the risk of reoperation was greater in those undergoing the primary operation within the early (five) years of the diagnosis of their Crohn's disease.

Steinberg JC, Allan RN, Brooke BN, Cooke WT, Alexander-Williams J: Sequelae of colectomy and ileostomy: Comparison between Crohn's colitis and ulcerative colitis. *Gastroenterology* 1975;68:33–39.

The carefully assembled results of a specially interested and competent group which supports the generally held view that Crohn's disease and ulcerative colitis behave differently after operation, and that patients with Crohn's disease run the risk of recurrences even with ileostomy.

Chapter 23. The Quality of Life and Inflammatory Bowel Disease

Moody GA, Mayberry JF: Quality of life: Its assessment in gastroenterology. *Euro J Gastro Hepatol* 1992;4:1025–1030.

One of the many papers which stress the definite need for a uniformly agreed, reliable value assessment to measure the quality of life and hence quality of care in gastroenterology.

Sickness Impact Profiles

Bergner M, Babbitt RA, Carter WB: The Sickness Impact Profile: Development and final revision of a health status measure. *Med Care* 1981;19:787–805.

One of several currently available quantitative validated measures of the impact of sickness on patients, adaptable to IBD.

Drossman DA, Lesserman J, Li Z, Mitchell M, Zaganis EA, Patrick DL: The Rating Form of IBD Patient Concerns: A new measure of health status. *Psychosomatic Medicine* 1991;53:701–712.

A contribution from the leading group at present concerned with measurement of quality of life in IBD with emphasis on the Rating Form of IBD Patient Concerns (RFIPC).

Love TJ, Irvine EJ, Fedoras RN: Quality of life in inflammatory bowel disease. *J Clin Gastro* 1992;14:15–19.

Using their own validated questionnaire the authors demonstrate, not surprisingly, that IBD adversely affects the quality of life in a highly motivated group of "well" outpatients when compared to an age- and sex-matched population.

Chapter 25. Pregnancy and Inflammatory Bowel Disease

Crohn BB, Yarnis H, Crohn E, Walter RI, Gabrilve LJ: Ulcerative colitis and pregnancy. *Gastroenterology* 1956;30:391–403.

Crohn BB, Yarnis H, Korelitz BI: Regional ileitis complicating pregnancy. *Gastroenterology* 1956;31:615–628.

Pioneer papers by pioneers in the field. Well worth rereading to measure the distance we have come in managing pregnancy and IBD.

Mogadam M, Dobbins WO, Korelitz BI: Pregnancy in inflammatory bowel disease: The effects of sulfasalazine and corticosteroids in fetal outcome. *Gastroenterology* 1981;80:72–76.

This study with a related one by the same authors (*Gastroenterology* 1980;78[part 2]:1224) emphasizes the safety for patient and fetus of the standard drugs (steroids and sulfasalazine) during pregnancy.

Mogadam M, Korelitz BI, Ahmed S, Dobbins WO: The course of inflammatory bowel disease during pregnancy and postpartum. *Am J Gastro* 1981;75:265–269.

Further information relating the likelihood of developing an exacerbation of IBD during and following pregnancy in the postpartum period.

Korelitz BI: Pregnancy, fertility and inflammatory bowel disease. *Am J Gastro* 1985;80:365–370.

A masterly summary of personal experience by a leader in this area.

Porter RJ, Stewart GM: The effects of inflammatory bowel disease on pregnancy: A case-controlled retrospective analysis. *Br J Obs & Gyn* 1986;93:1124–1131.

A reassuringly nice study from the obstetrical point of view on the safety of pregnancy in IBD patients with the conclusion that "in the absence of a relapse, a diagnosis of inflammatory bowel disease should not influence obstetric management."

Singly AJ, Brandt LJ: Pathophysiology of the gastrointestinal tract during pregnancy. *Am J Gastro* 1991;86:1695–1712.

A comprehensive review of the alterations in physiology of the entire gastrointestinal tract during pregnancy.

Korelitz BI: Inflammatory bowel disease in pregnancy. *Gastroenterology Clinics of North America* 1992;20:827–833.

The latest but not last we hope word on this subject by the pioneer in this area who continues his clinical observation on the front line.

Chapter 26. The Cancer Problem

Ulcerative Colitis

DYSPLASIA

Morson BC, Pang LSC: Rectal biopsy as an aid to cancer control in ulcerative colitis. *Gut* 1967;8:423–434.

The classic paper that presented the evidence linking dysplasia to cancer in ulcerative colitis—a landmark.

CANCER

Devroede GJ, Taylor WF, Sauer WG, Jackman RJ, Stickler GB: Cancer risk and life expectancy in children with ulcerative colitis. *N Engl J Med* 1971;285:17–21.

The exemplary illustration of the use of life tables (actuarial methods) for the problem of the development of cancer in ulcerative colitis as a function of duration of illness.

Lennard-Jones JE, Morson BC, Ritchie JK, Shore DC, William CB: Cancer in colitis: Assessment of the individual risk by clinical and histologic criteria. *Gastroenterology* 1977;73:1280–1289.

A classic study, prospective in nature, in which the risk to an individual was assessed by careful follow up, sigmoidoscopy and biopsy for dysplasia and which was instrumental in putting this problem in its present perspective.

Nugent FW, Haggi HRC: Results of a long-term prospective surveillance program for dysplasia in ulcerative colitis. *Gastroenterology* 1984; 80(part 2):1197.

An example of the current prospective approach to surveillance in ulcerative colitis based on colonoscopic biopsy for dysplasia.

Itskowitz S: Colorectal cancer in inflammatory bowel disease. *Progress in Inflammatory Bowel Disease* 1993;14:1–5.

A lucid and succinct account of current information regarding risk factors in ulcerative colitis–cancer, the biology of clinical and molecular events in the adenoma–carcinoma sequence, including a discussion of dysplasia and aneuploidy, and newer tumor markers such as the mucin associated antigen.

Crohn's Disease

Ginzburg L, Schneider KM, Dreizin DH, Levinson C: Carcinoma of the jejunum occurring in a case of regional enteritis. *Surgery* 1956;39:347–351.

The first small bowel cancer associated with small bowel Crohn's disease by one of the original discoverers of regional enteritis.

Weedon DD, Shorter RG, Ilstrup DM, Huizenga KA, Taylor WF: Crohn's disease and cancer. *N Engl J Med* 1973;289:1099–1102.

The first well-followed up group of a large number (449) patients with Crohn's disease studied by life table methods, which demonstrated the incidence of colorectal cancer to be greater than in a control population. An astonishing number of patients were followed up, 442 cases (92.4%).

Gyde SN, Prior P, Macartney JC, Thompson H, Waterhouse JHH, Allan RN: Malignancy in Crohn's disease. *Gut* 1980; 21: 1024–1029.

A further study of 513 cases of Crohn's disease followed from 1944 to 1976, again demonstrating an increased risk of cancer, relative risk 1.7 at all sites, and 3.3 for cancer of the gastrointestinal tract, both upper and lower. The situation seems to resemble that in gluten enteropathy, for tumors in the upper gut.

Seney E, Sachar DB, Keshane M, Greenstein AJ: Small bowel carcinoma in Crohn's disease: Distinguishing features and risk factors. *Cancer* 1989; 63:360–363.

Yamazaki Y, Ribeiro B, Sachar DB, Amper AM, Greenstein HJ: Malignant colorectal stricture in Crohn's disease. *Am J Gastro* 1992;86:882.

Ekborn A, Helmick SH, Zack M, Adami O: Increased risk of large bowel cancer in Crohn's disease with colonic involvement. *The Lancet* 1990;336:357–359.

More recent reports reinforcing the evidence of the risk of cancer in Crohn's disease which strongly confirm that this risk is equal to that in ulcerative colitis.

Chapter 27. Associated Diseases

Autoimmune Diseases

Snook JA, da Silva HJ, Jewell PJ: The association of autoimmune disorders with inflammatory bowel disease. *Quart J Med* 1989;72:835–840.

A scholarly study of the increased incidence of autoimmune disorders in patients with ulcerative colitis in contrast to outpatient controls and patients with Crohn's disease.

Psoriasis

Yates VM, Watkinson G, Kelman H: Further evidence for an association between psoriasis, Crohn's disease and ulcerative colitis. *Brit J Dermatol* 1982;106:323–330.

The prevalence of psoriasis in Crohn's disease (11%) and in ulcerative colitis (5.7%) was greater than in a control group (1.5%). Psoriasis was also increased in first-degree relatives with IBD. The authors suggest a relation between psoriasis, ankylosing spondylitis, sacroiilitis, and IBD through HLA B27.

Neutrophilic Disorders

Vannier JP, Araud-Battandier F, Riccour C, et al: Chronic neutropenia and Crohn's disease of childhood. Report of two cases. *Arch Fr Pediatr* 1982;39:367–370.

Roe TF, Thomas DW, Citranz V, et al: Inflammatory bowel disease in glycogen storage disease Ib. *J Pediatr* 1986;109:55–59.

Stevens C, Peppercorn MA, Grand RJ: Crohn's disease associated with autoimmune neutropenia. *J Clin Gastro* 1991;13:328–330.
An interesting group of patients with IBD and disorders of neutrophilactivity.

Chapter 28. Extraintestinal Manifestations

General

Kern, F Jr: Extraintestinal manifestations. In Kirsner JB, Shorter RG (eds): *Inflammatory Bowel Disease.* Philadelphia, Lea & Febiger, 1980, pp 217–240; 470–473.

Kidney Stones

Deren JJ, Porush JG, Levitt MF, Khilnani MT: Nephrolithiasis as a complication of ulcerative colitis and regional enteritis. *Ann Intern Med* 1962;56:843–853.
An early review of the Mount Sinai Hospital experience of 538 patients with ulcerative colitis and regional ileitis, with 28 patients with stones (4.8%), emphasizing uric acid and calcium oxalate stones, with some perspicacious anticipation of current thinking on volume depletion and urine pH.
Smith LH, Fromm H, Hoffman HF: Acquired hyperoxaluria, nephrolithiasis, and intestinal disease. Description of a syndrome. *N Engl J Med* 1972;286:1371–1375.
A classic description of this syndrome and the use of cholestyramine as therapy for hyperoxaluria.

Liver

Eade NN: Liver disease in ulcerative colitis. I. Analysis of operative liver biopsy in 138 consecutive patients having colectomy. *Ann Intern Med* 1970;723:475–487.
Eade NN, Cooke WT, William JA: Liver disease in Crohn's disease: A study of 100 consecutive patients. *Scand J Gastro* 1971; 6:194–204.
A large series of biopsies in both ulcerative colitis and Crohn's disease indicating the variety of lesions found routinely.

Tremaine WJ: The liver and inflammatory bowel disease: An update for the '90s. *Progress in Inflammatory Bowel Disease* 1992;13:1–4.
A lucid and succinct account of liver disease in IBD.

Joints

Keat A: Reiter's syndrome and reactive arthritis in perspective. *N Engl J Med* 1983;309:1606–1615.
An illuminating review from the rheumatologist's perspective of reactive arthritis with a fine bibliography on the joints problem associated with intestinal disease.

An Album of Small Bowel Radiographs

Errors in differential diagnosis, I believe, arise mainly from the failure to spread the diagnostic net wide enough. We fail to include a large enough list of possibilities. One of the commonest errors I have found is the overwhelming tendency to label every radiograph in which there are abnormalities in the ileum as being Crohn's disease. The following album is designed to stress, perhaps overstress, this point.

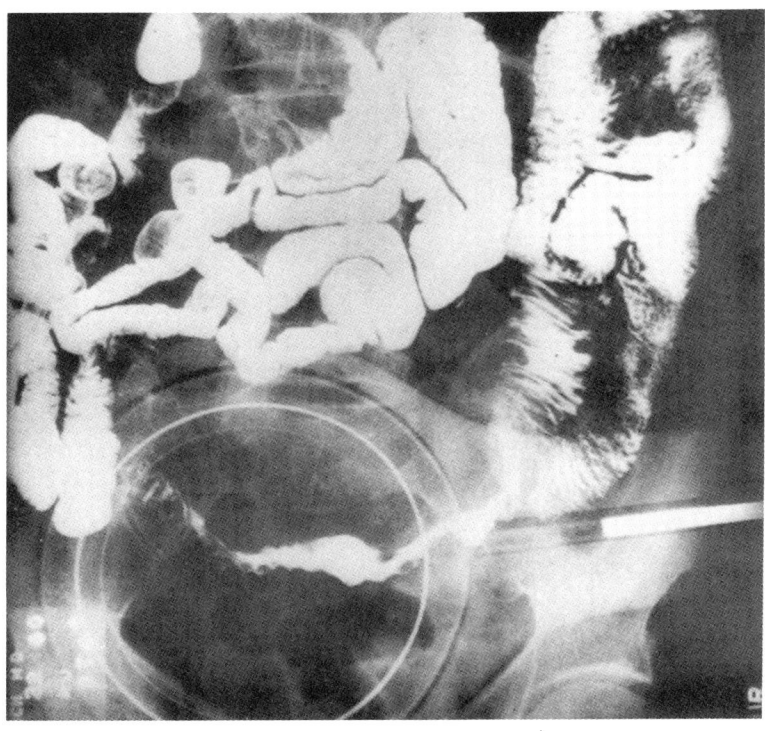

FIGURE 1 Plasmacytoma of ileum (not part of generalized multiple myeloma)

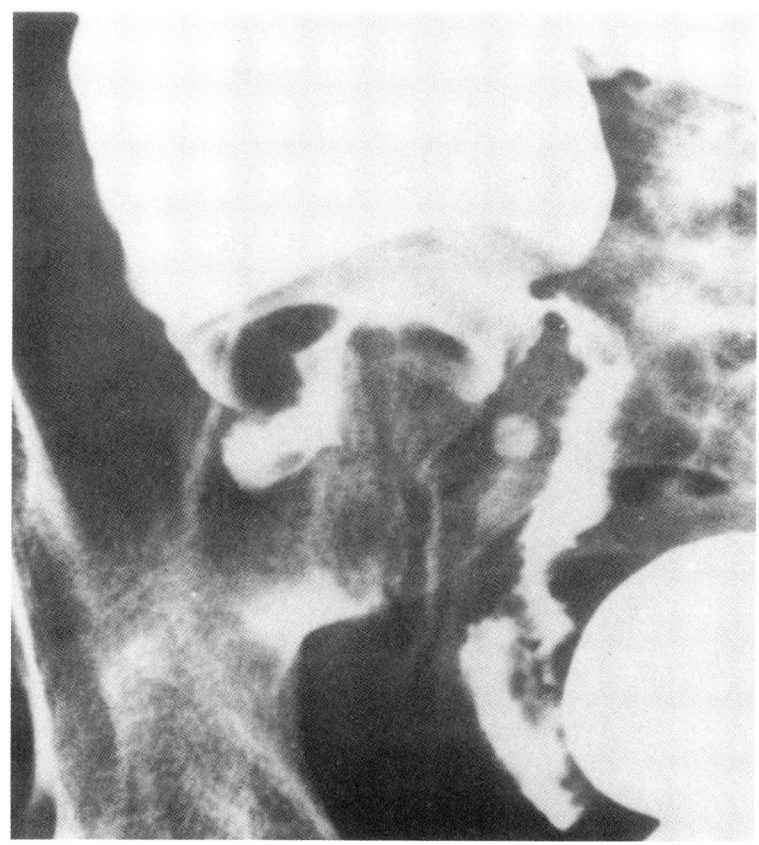

FIGURE 2 Appendiceal abscess, one of the commonest conditions confused with Crohn's disease

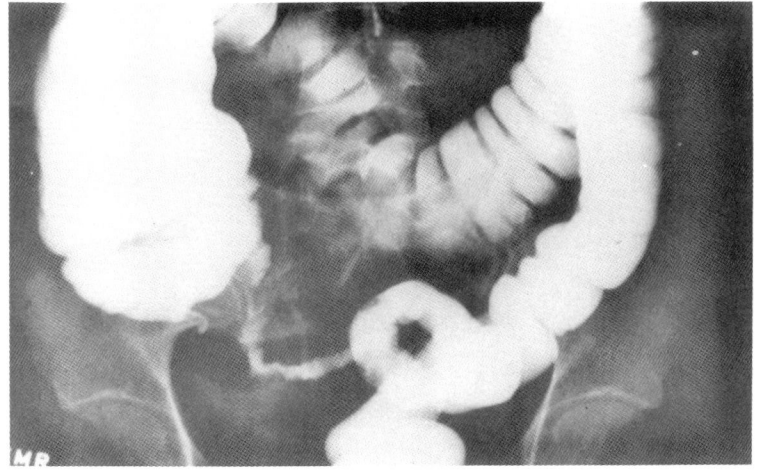

FIGURE 3 Radiation enteritis, less frequent than radiation colitis

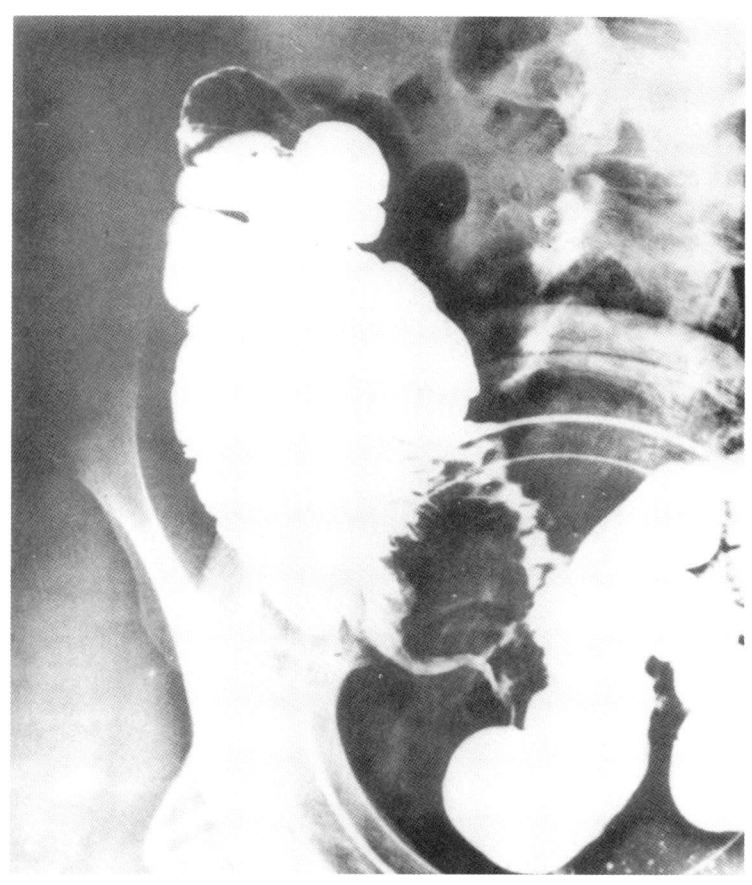

FIGURE 4 Intramural bleeding in hemophilia

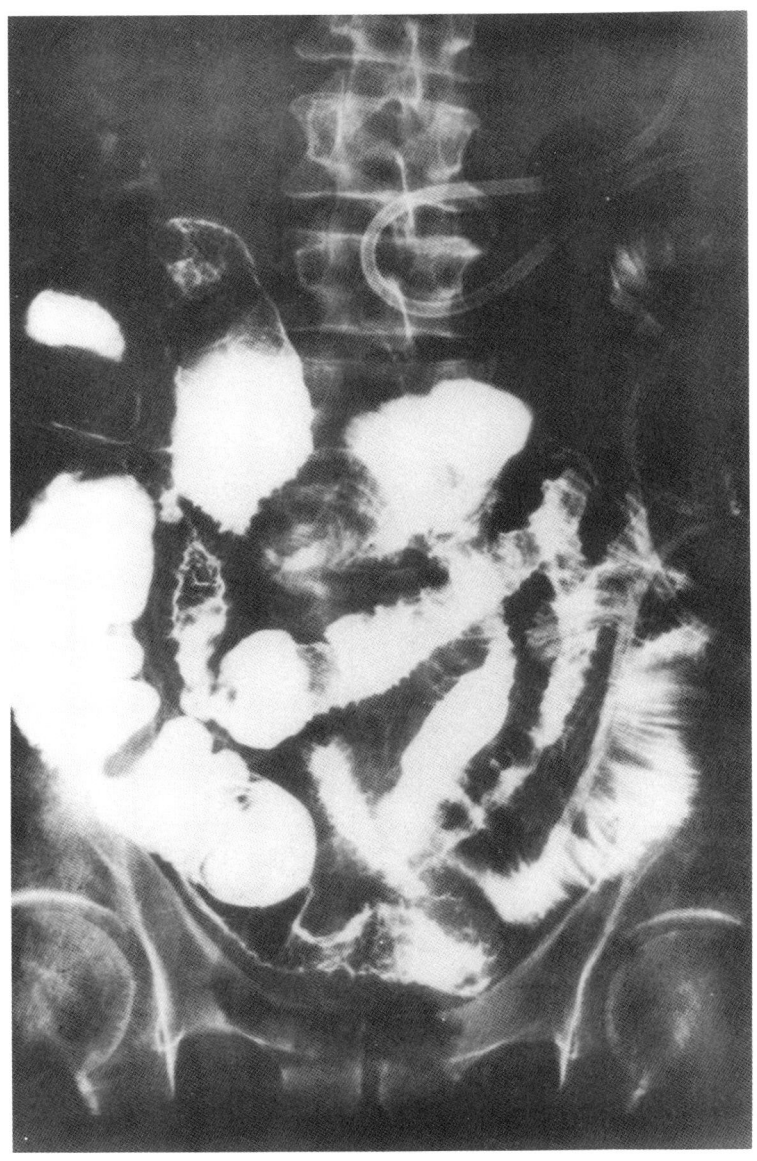

FIGURE 5 Tuberculosis, rare at present

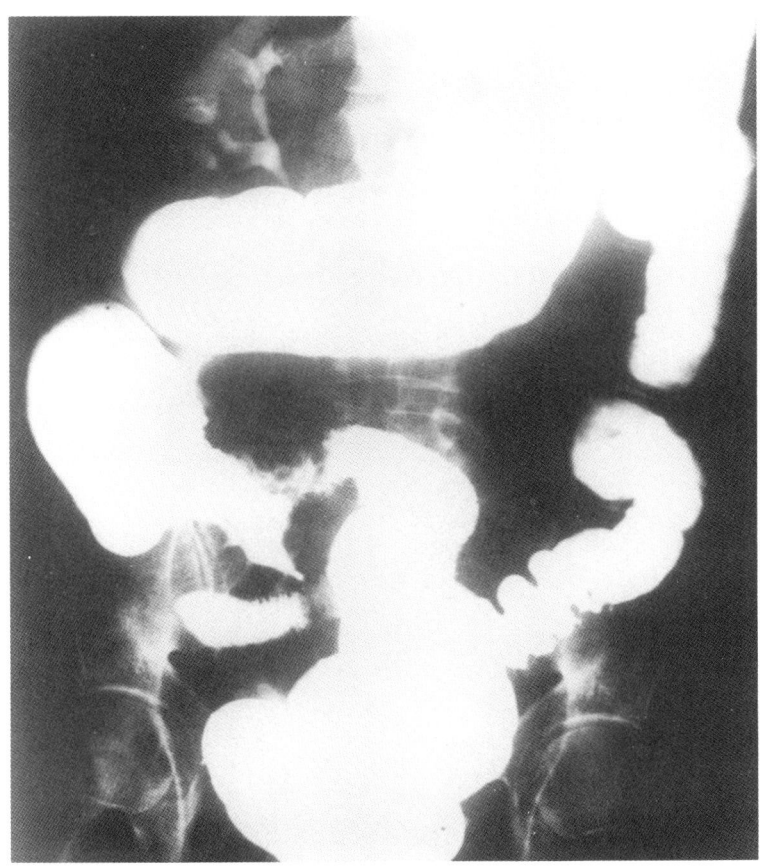

FIGURE 6 Tuberculosis

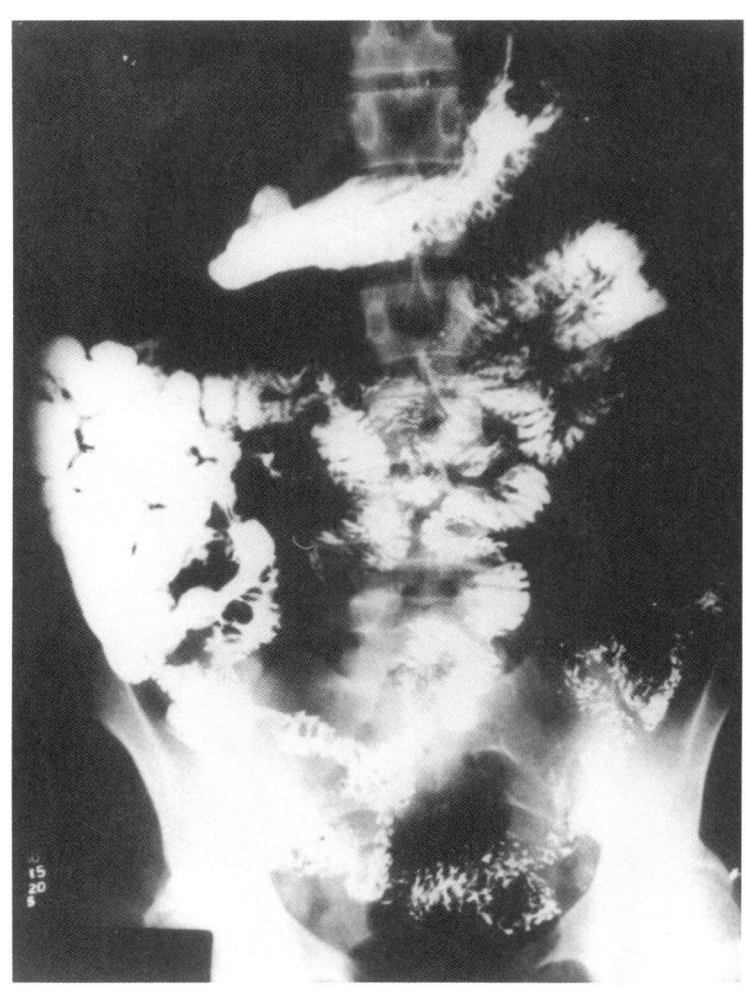

FIGURE 7 Lymphangioma

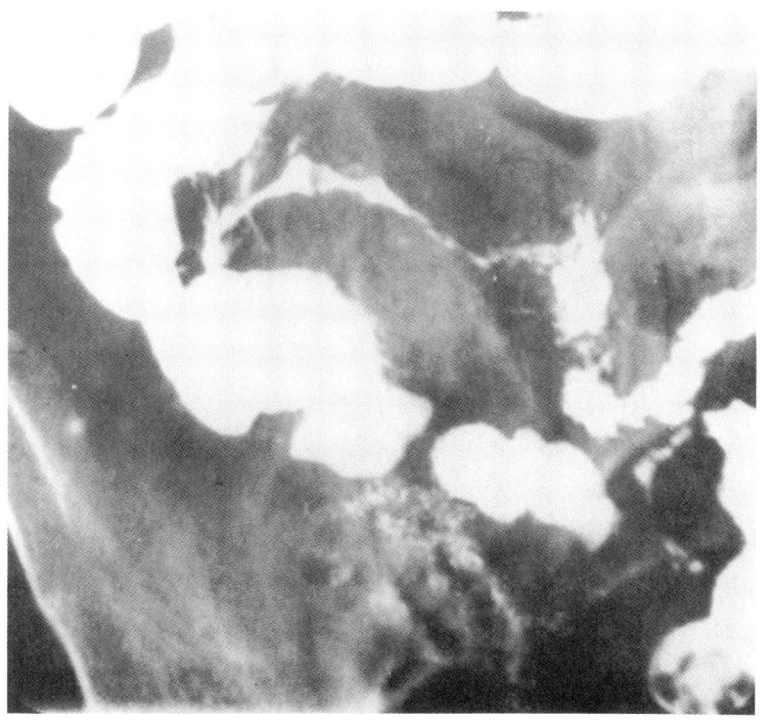

FIGURE 8 Carcinoid tumor, nonfunctional

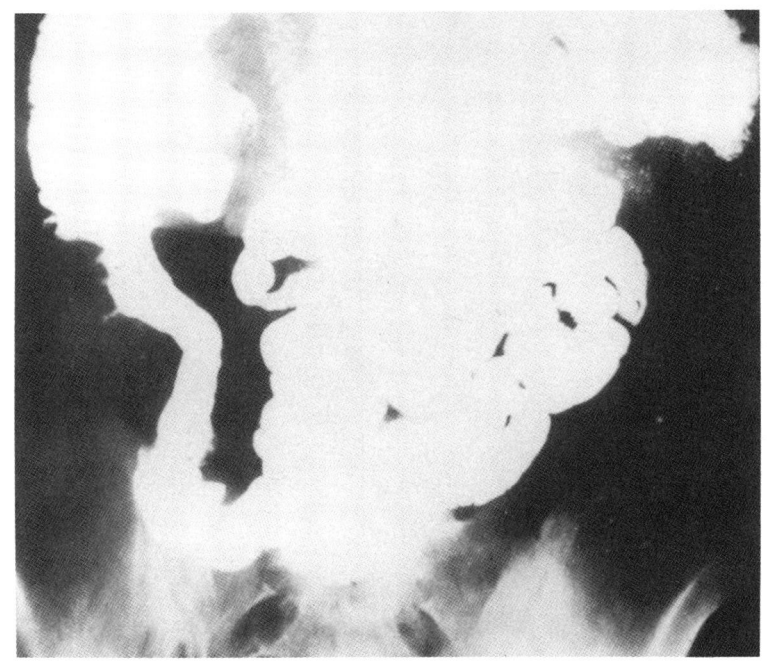

FIGURE 9 Carcinoid tumor

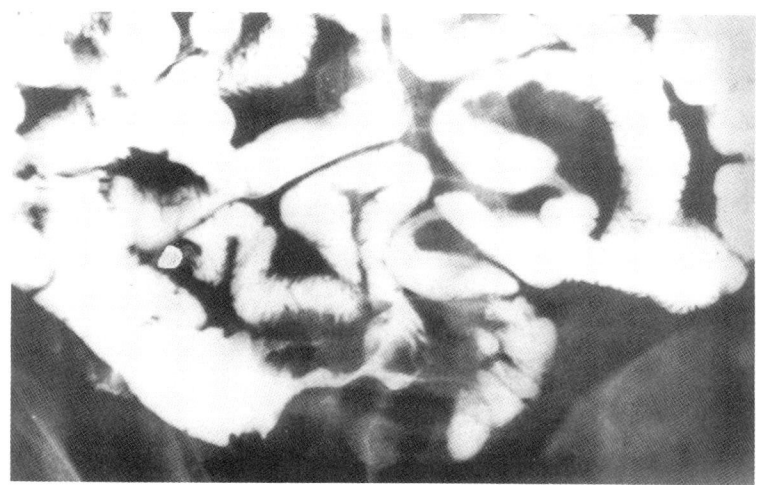

FIGURE 10 Hodgkin's disease

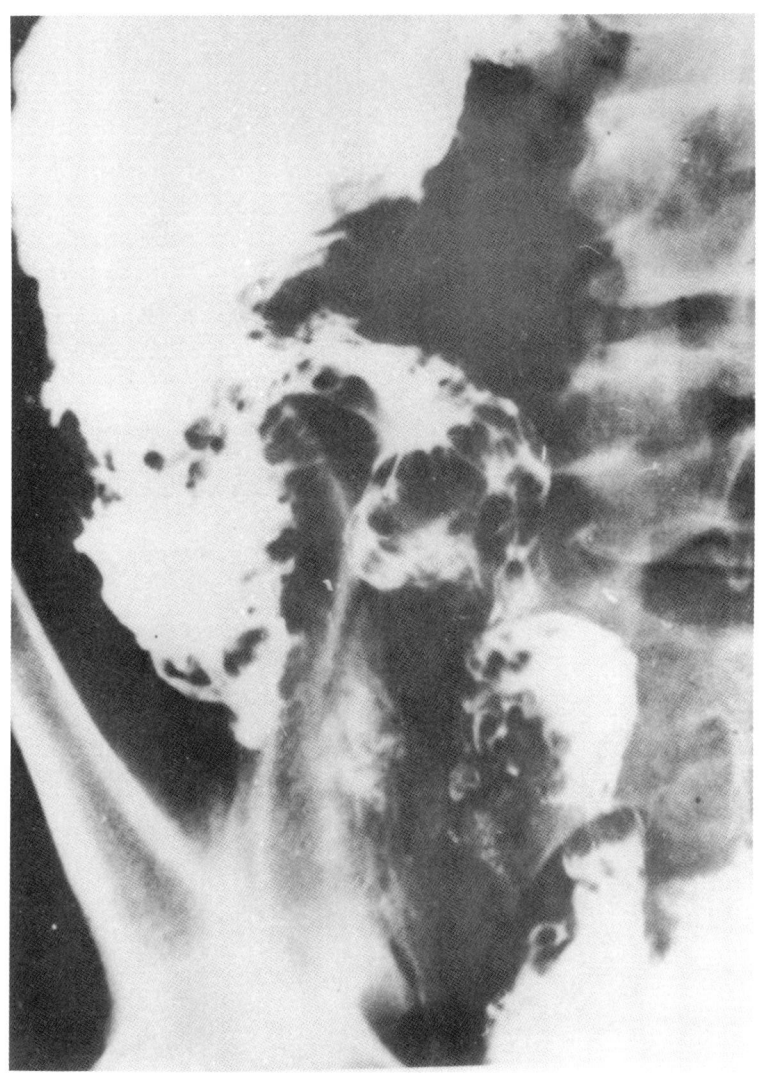

FIGURE 11 Lymphosarcoma invading the cecum

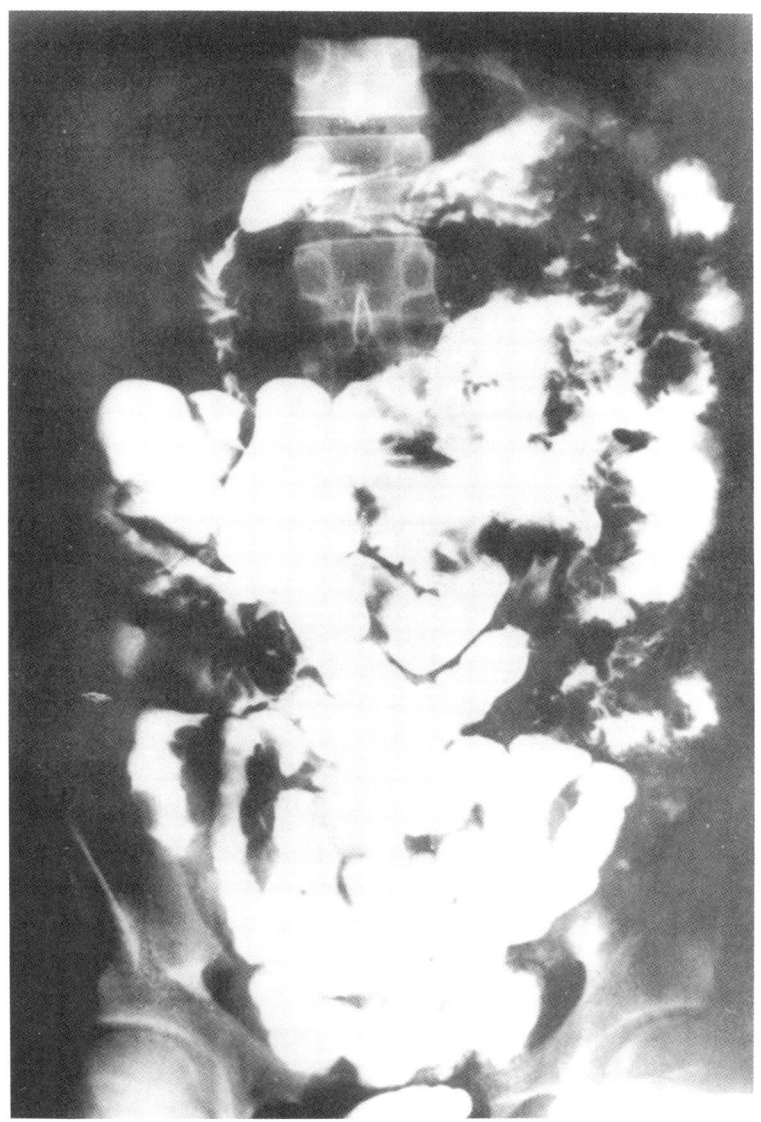

FIGURE 12 Diffuse lymphosarcoma

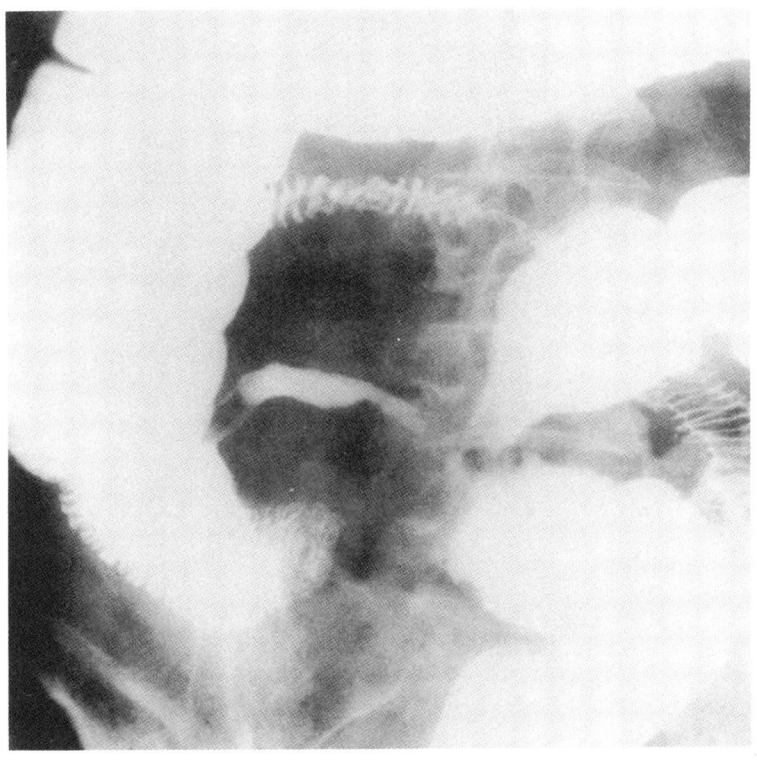

FIGURE 13 Carcinoma of the cecum can on occasion invade the ileum

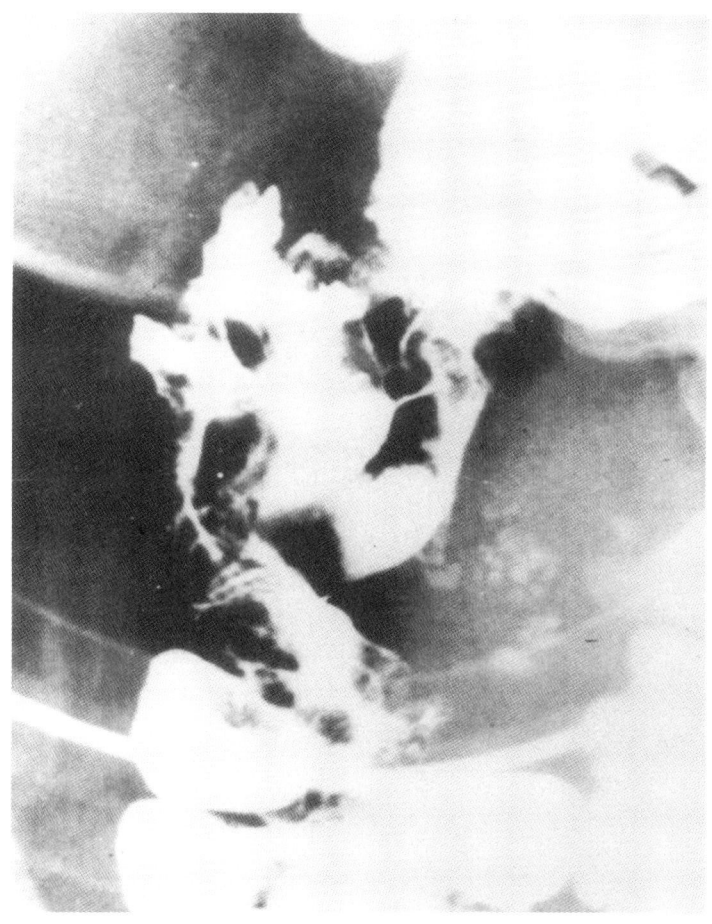

FIGURE 14 Carcinoma of the cecum

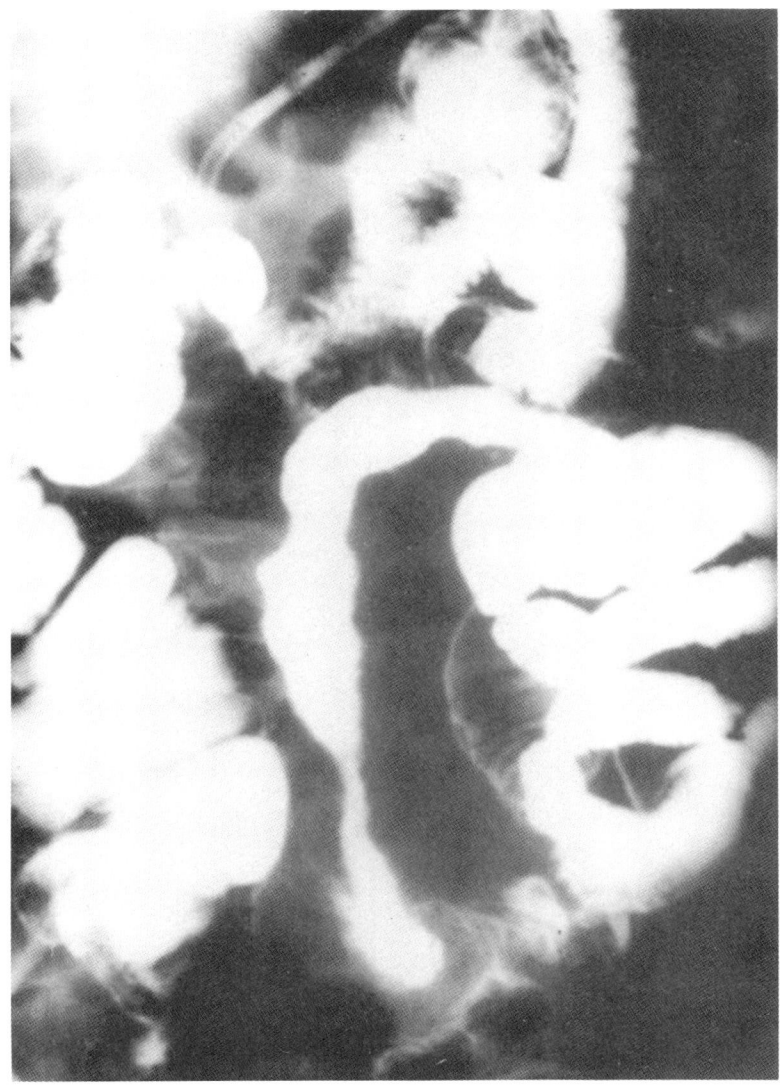

FIGURE 15 Scirrhous carcinoma of the ileum, not related to Crohn's disease

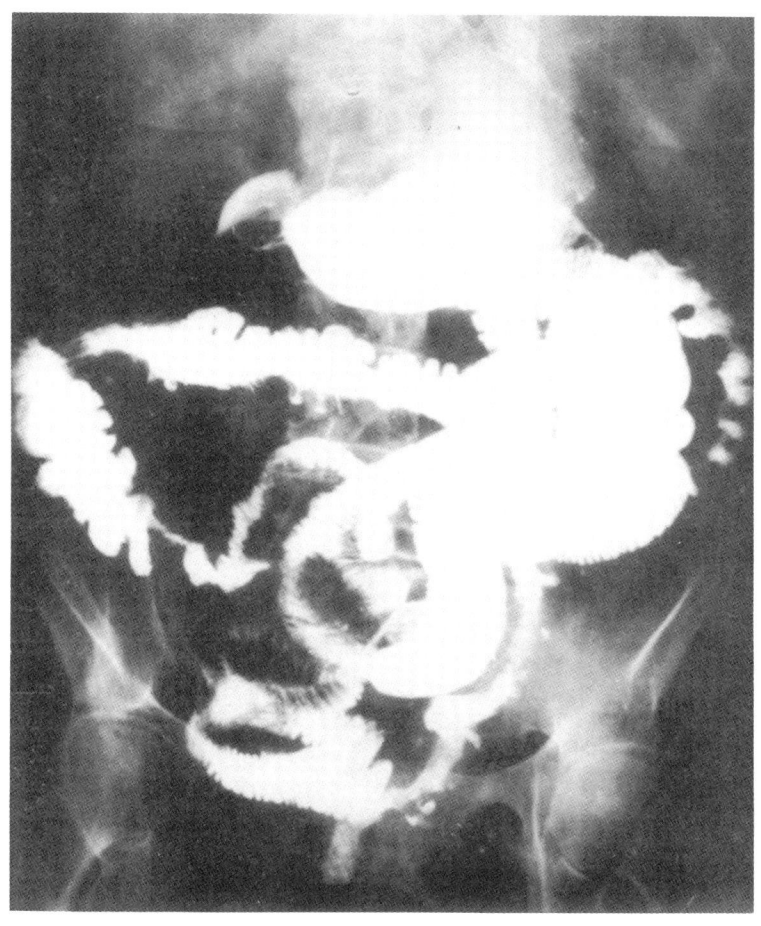

FIGURE 16 Metastatic implants from breast cancer

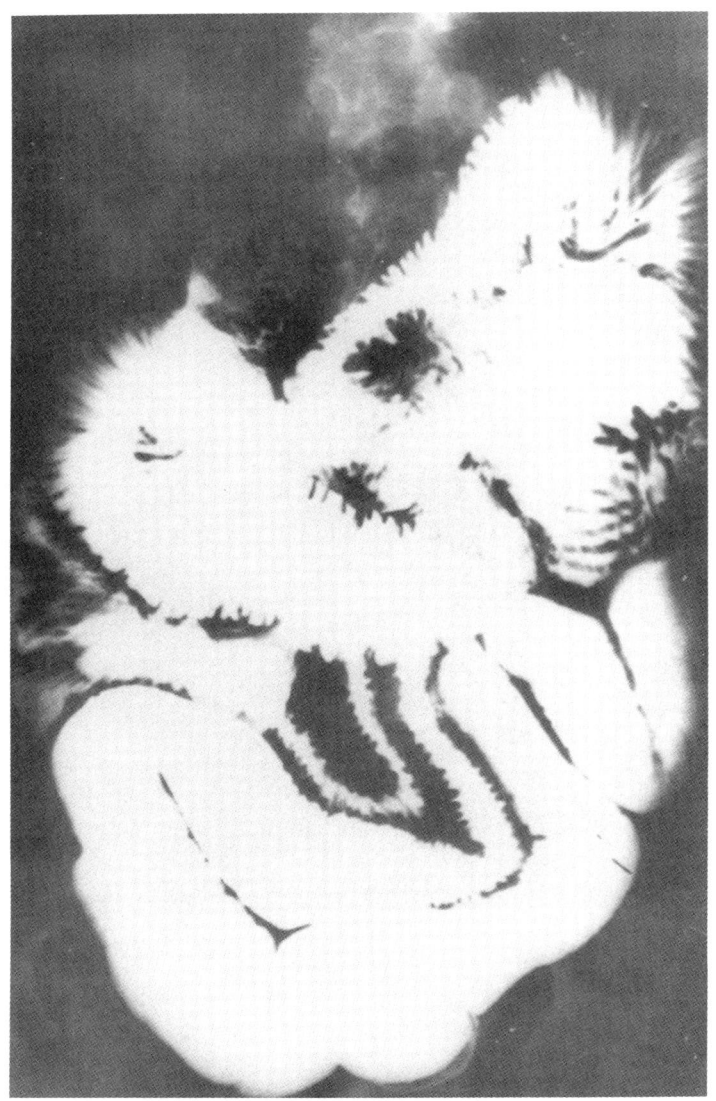

Figure 17 Pseudo-obstruction

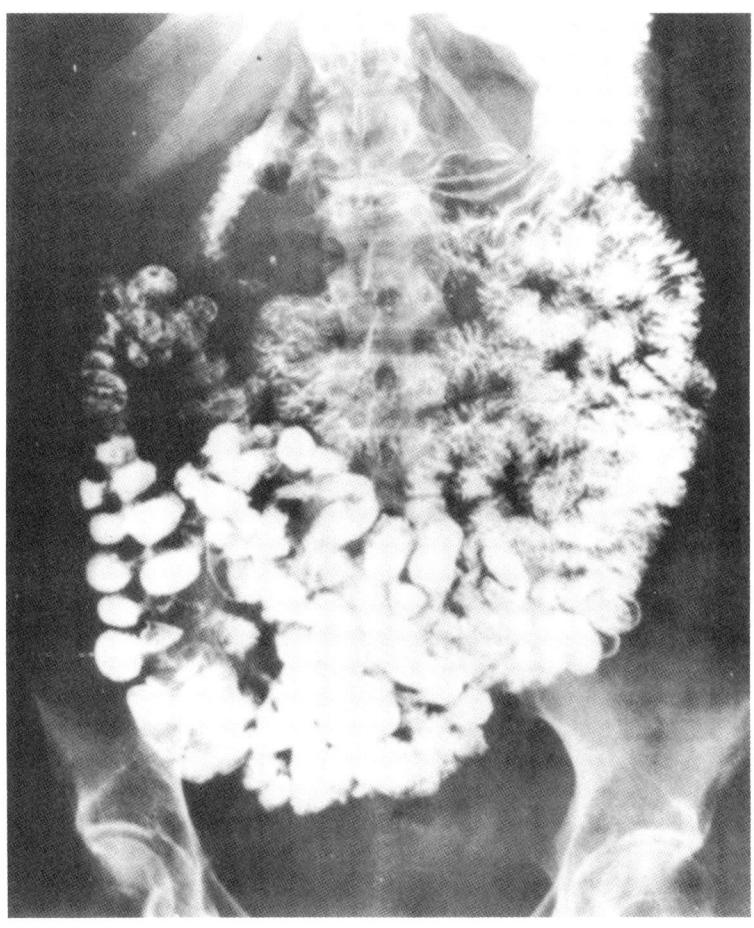

FIGURE 18 Lupus vasculitis

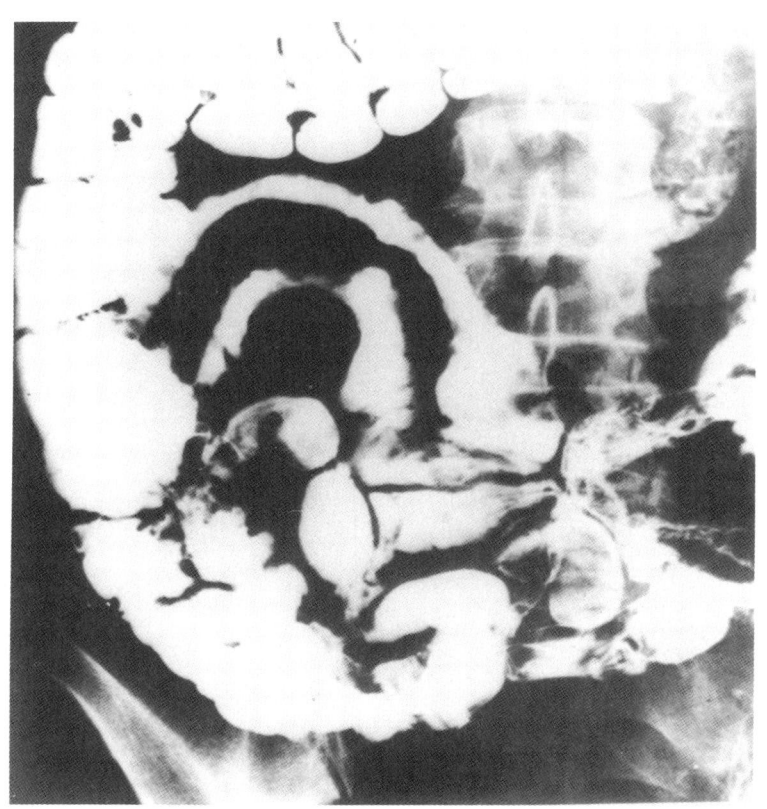
FIGURE 19 Ischemia of small bowel

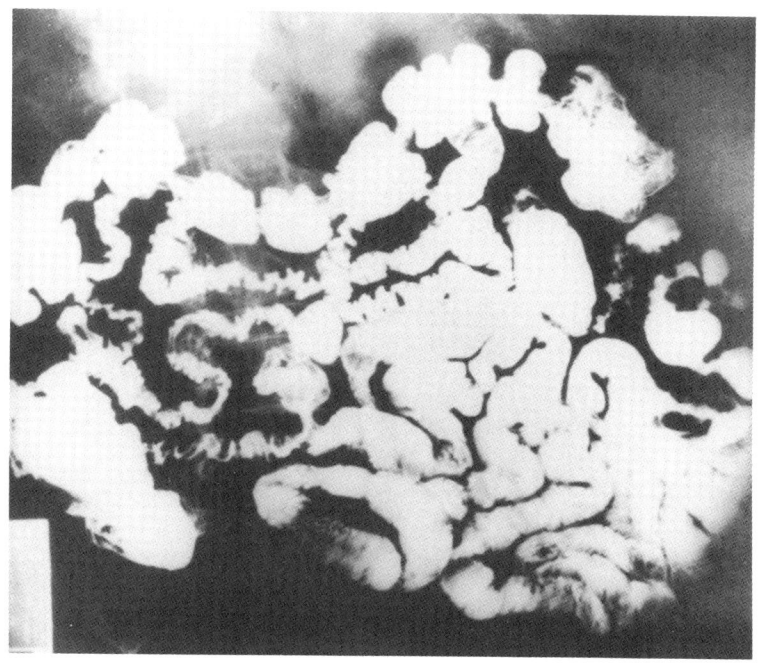

FIGURE 20 Ischemia of small bowel, secondary to contraceptive pill

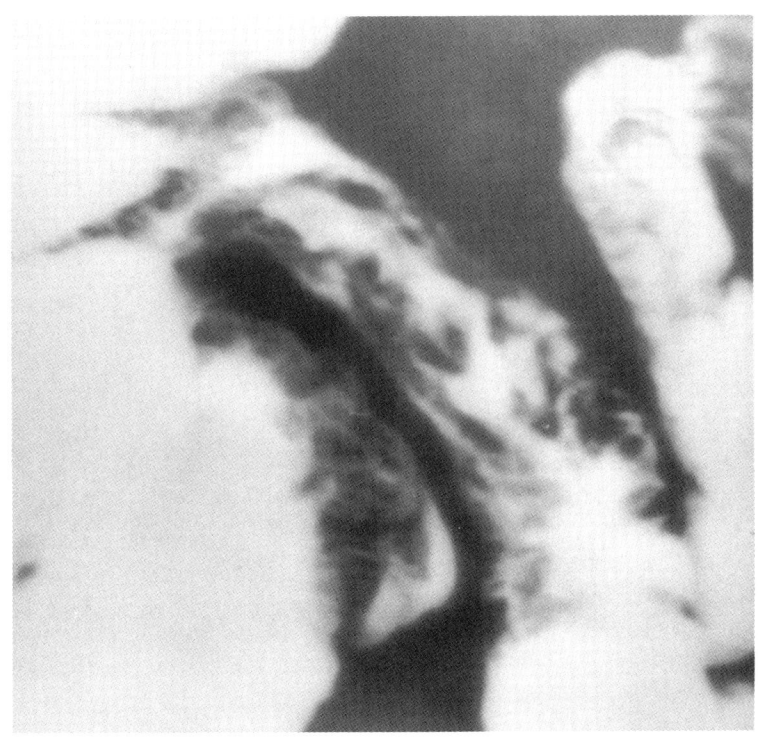

FIGURE 21 Lymphoid hyperplasia often confused with regional ileitis

Index

Abdomen, physical examination, 52–53
Abdominal fistulas, 211–213
  sites of, 211–212
  surgery, 167–168
  variants, 212–213
Abdominal pain, assessment, 90–91
Abdominoperineal resection, 194
Abscess
  appendiceal, 74
    radiograph, 295
  in Crohn's disease, 178
  liver, colloid scanning in, 61
  perirectal, 214–215
  surgery indication, 167–168, 178
ACTH
  in Crohn's disease, 139
  in ulcerative colitis, 120, 128–130
Acute autoimmune hemolytic anemia, 234
Acute myelogenous leukemia, 228–229, 232
Adenomatous polyps, 225–226
Adrenal hemorrhage, and ACTH, 120, 130
Age factors, and recurrence, 187
Albumin, 90–91
Alpha-1-anti-trypsin
  Crohn's disease, 90
  diagnostic utility, 93
5-Aminosalicylates
  available drugs, 114
  general considerations, 116–119
Amoebiasis, diagnosis, 56

Amyloid A-protein, 246–247
Amyloidosis, 243–246
  clinical presentation, 245
  colchicine in, 245–246
  treatment, 245–246
Anal squamous cell cancer, 228
Anastomosis. *See also specific types*
  cancer risk, 224
  and recurrence, Crohn's disease, 176, 187–188
Anerobic bacteria, 23–24
Ankylosing spondylitis, 237–238
Antacids, 130
Antibiotics
  available drugs, 115
  in Crohn's disease, 135–137
  general considerations, 121–122
  ileovesicular fistula treatment, 209
  in ulcerative colitis, 128
Antineutrophilic cytoplasmic antibody
  and refractory pouchitis, 67
  ulcerative colitis versus Crohn's disease, 10–11, 27
Antituberculosis drugs, 23
Anxiety, 152
Appendectomy scar, fistula, 212
Appendiceal abscess
  differential diagnosis, 74
  radiograph, 295
Appendicitis
  and Crohn's disease, 45–47, 212
  in inflammatory bowel disease, 45–47
  sonography, 46
  and ulcerative colitis, diagnosis, 69

315

Arachidonic acid metabolism, 24–25
Arthralgias, 237–238
  incidence, 237
  steroid induced, 131
Arthritis, 237–238
  colectomy effect on, 165
  and diversionary colitis, 197
5-ASA
  anti-inflammatory mechanism, 25, 27
  in Crohn's disease, 134–135, 142
  extraintestinal side effects, 249
  maintenance therapy value, 27, 113, 132–133, 142
  oral preparations, 118–119
  and pregnancy, 219
  in proctitis, 125–128
  in severe ulcerative colitis, 130
  in universal ulcerative colitis, 128
Asacol®, 118, 133
Aseptic bone necrosis, 132, 249
Asthma, 231
Atabrine, 82
Atonic bladder, 243
Atopic disorders, 231
Autoimmune disorders, 233–234
  Crohn's disease versus ulcerative colitis, 10
  incidence, 234
Azathioprine
  in Crohn's disease, 140–142
  cutaneous fistula treatment, 213
  extraintestinal side effects, 250
  general considerations, 122–123
  maintenance use, 133, 142
  meta-analysis, 156
  and pregnancy, 219
  and steroid weaning, 131
  in ulcerative colitis, 133
Azulfidine®. See Sulfasalazine

Balsalazide, 133
Barium enema
  contraindications, 57
  in Crohn's disease, 60
  in ulcerative colitis, 58–59
Beta-methasone-17-malonate, 120
Bile-salt catharsis, 203, 241
Bile-salt excretion, 92, 241
Biliary tract disease, 241–242
Biopsy, 58–59

Bladder fistula
  diagnosis and treatment, 208–209
  and massive hematuria, 42
Bladder manifestations, 246
Bleeding
  Crohn's disease atypical presentation, 39–40, 42
  emergency surgery decisions, 162, 166
  labeled red cell scan, 61
  and ulcerative proctitis, 32–33
  universal ulcerative colitis, 33–34
Breast cancer metastasis, radiograph, 309
Brooke ileostomy
  in Crohn's disease of colon, 178
  and pancolectomy, 173–174
  postoperative intestinal obstruction, 192
  results, 181–182
Budesonide, 25, 121
Bypass surgery
  and cancer, 226–227
  versus resection, Crohn's disease, 176–177

C-reactive protein, 89–90
Calcium supplements, 146
Campylobacter infections, 73
Cancer, 222–229
  and colectomy decision, 163
  and Crohn's disease, 226–229
    risk, 226
    surgery, 167
    versus Crohn's ileitis, 73–74
  diagnostic errors, 75–77
  differential diagnosis, 73–74, 77
  doubling time, 225
  proctitis risk, 97–98
  radiographs, 300–309
  and ulcerative colitis, 222–224
    diagnostic errors, 75–76, 223–224
    endoscopy timing, 59
    and ileorectal anastomosis, 185, 193
    proctitis risk, 97–98
    prognosis, 224
    risk, 222–223
    subtotal colectomy risk, 192–193
Cancer surveillance, 224–226
Carcinoid tumor, radiographs, 301, 302

CD4+ cells, 26
CD8+ cells, 26
Cecal cancer
 differential diagnosis, 73, 77
 radiographs, 306, 307
Cerebrovascular accidents, 249
Cesarean section incision, fistula, 212
Chest examination, 53
Cholangiosarcoma, 247
Cholestasis, 247
Choroiditis, 237
Ciprofloxacin, 136
Circulating immune complexes, 93–94, 236
Clofazamine, 23
Clostridia difficile toxin, 36
Colazide®, 133
Colchicine, 245–246
Colectomy
 Crohn's colitis complications, 194
 decision for, 162–165
 elective, 164–165
 emergency decisions, 162
 predictors of, 96
 prophylactic, 163–164, 225
 results, 181–182
Collagenous colitis
 collagenous sprue relationship, 82
 diagnostic errors, 77, 81–82
 versus microscopic colitis, 83
 in postmenopausal women, 81
Collagenous sprue, 82
Colon. *See also* Toxic dilation of colon
 physical examination, 53
Colon cancer
 and colectomy decision, 163
 and Crohn's disease, 228
 surgery results, 182
Colonic fistulas, 210–211
Colonic mucosal prolapse, 81
Colonoscopy
 in cancer surveillance, 225
 in Crohn's disease, 60, 70
 in ulcerative colitis, 58–59
Computed tomography, 61–62
Congenital neutropenia, 233
Conjunctivitis
 initial presentation, 35, 236–237
 physical examination, 51
Contagion, 20

Contour defect, 57
Contraceptive pill, 74–75
 secondary ischemia, radiograph, 313
Corticosteroids. *See* Steroids
Corticotropin. *See* ACTH
Cortifoam®, 126–127
Cortisone derivatives, 119–121
"Crohn tags," 213
Crohn's and Colitis Foundation of America, 174, 181
Crohn's colitis
 endoscopy, 72
 recurrence, 194–195
 surgery complications, 194–195
Crohn's disease
 antibiotics in, 135–137
 atypical presentations, 39–40
 clinical presentation, 37–47, 89–91
 of colon, fistulas, 210–211
 diagnostic difficulties, 69–71
 differential diagnosis, 72–79, 84
 drug strategies, 134–143, 154–158
 emergency surgery decisions, 166
 endoscopy, 60
 epidemiologic trends, 13
 etiologic theories, treatment implications, 19–27
 genetic factors, 12–13, 20–22
 immunologic theories, and treatment, 25–27
 infectious agents, 22–24
 laboratory studies, 89–92
 maintenance therapy, 141–142
 natural history of, 11
 pathology, 11–12
 perforating versus nonperforating, 47, 101–102
  prognostic implications, 101–102
 prognosis, 100–102, 110–111
 psychological factors, 152
 quality of life, 202–203
 scanning techniques, 61–62
 silent, 40–41
 steroid weaning, 138–139
 subgroups, 47
 suppurative complications, surgery, 167–168
 surgery, 165–169, 172–179
  of colon, 172–178
  complications, 193–199
  decision for, 165–169

Crohn's disease: surgery (*Cont.*)
  elective, 168–169
  results, 186–188, 202–203
  ulcerative colitis distinction, 8–14, 65–67
  vascular pathogenesis hypothesis, 248
Crohn's Disease Activity Index, 87–88, 92
Cryoglobulins, 249
Cutaneous fistulas, 167–168, 211–213
  multiple form, 213
Cyanosis, 248–249
Cyclosporine
  in Crohn's disease, 141–142
  helper T-cell inhibition, 26
  indications, 123
  maintenance use, 133, 142
  in perirectal abscess, 215
  and pregnancy, 221
  in severe ulcerative colitis, 129
  side effects, 123

Declomethasone dipropionate, 120
Dependency, 152
Depression, 152
Desmoplastic reactions, 84
Diagnosis, 56–93
  differential, 72–84
  difficulties of, 68–71
  errors in, 75–79
  imaging techniques, 56–62
  limits of, 65–67
Diarrheal disease
  collagenous colitis, 81
  and end ileostomy, 196
  following ileocecal resection, 194
  initial investigation, 55
  and microscopic colitis, 83
Diet, 144–149
  as etiological theory, 22, 144–145
  supplements, 145–147
Diet therapy, 145–148
Dietary fiber, 147–148
Dipentum®, 118, 126, 133
Diversionary colitis, 196–197
Diverticular disease, 75, 78–79
Dizygotic twins, 21
Doubling time, cancer, 225
Drug therapy, 112–143
  acute versus maintenance, 112–113

  in Crohn's disease, 134–143
  extraintestinal side effects, 249–250
  general considerations, 112–124
  meta-analysis, 154–158
  in ulcerative colitis, 125–133
Duodenal Crohn's disease, 44–45
Dysplasia
  and cancer surveillance, 225
  colectomy indication, 164
  and ulcerative colitis diagnosis, 75–76

Elective surgery, 164–165, 168–169
Elemental diets, 149
"Elephant ears," 213
Emergency decisions, surgery, 162
Emotional distress, 36–37, 150–153
End ileostomies
  complications, 195–196
  cutaneous fistula, 212
Endoscopic retrograde cholangio-pancreatogram, 247
Endoscopy, 57–60
  in Crohn's colitis, 92
  in Crohn's disease, 60
  as recurrence evidence, 186–187
  in ulcerative colitis, 58–59
Entamoeba histolytica, 75
Enteral nutritional therapy, 149
Enteroenteric fistulas, 168, 207–209
Enterovesicular fistulas, 168
Epidemiology, 13, 19
Erythema multiforme, 249
Erythema nodosum, 35, 239
Erythrocyte sedimentation rate
  Crohn's disease, 89
  problems in small-bowel disease, 89
Esophageal Crohn's disease, 44–45
Esophageal fistulas, 215–216
Ethambutol, 23
Eudraget S, 121
Extraintestinal manifestations, 235–251
  drug therapy complications, 249
  factitious form, 250–251
  initial presentation, 34–35
  and pouchitis, 198
Extremities, physical examination, 52
Eye manifestations, 51, 236–237

Factitious disease, 250–251
Familial colonic polyposis, 183
Familial Mediterranean Fever, 245
Family studies
  Crohn's disease versus ulcerative colitis, 12–13, 20–22
  and genetic risk, 218
Fat malabsorption, 91–92, 241
Fatty liver, 53
Fertility rate, 217
Fetal wastage, 218–219
Fever, 190–191
Fiber-enriched diets, 148
Fingers
  gangrene, 77, 248–249
  physical examination, 52
Fistulas, 207–216
  abdominal wall, surgery, 167–168
  colonic forms, 210–211
  cutaneous, abdominal wall, 211–213
  intestinal forms, 207–210
  perianal and perirectal, 213–215
  and pregnancy, 220
  stenosis in pathogenesis, 208
FK 506, 124
Flagyl®. See Metronidazole
Flexible sigmoidoscopy. See Sigmoidoscopy
Fluticasone, 120
Folate absorption, 117
Folate supplements, 134, 146
Free perforation
  and colectomy decision, 162
  in Crohn's disease, 38–39, 43–44, 101–102, 166
    surgery, 166
  in jejunum, 43–44
  prognostic implications, 101–102
  and ulcerative colitis, 34

Gallstones, 230–231, 241–242
Gangrene, 77, 248–249
Gastroduodenal Crohn's disease, 44–45
Genetics, 20–22
  etiological implications, 20–22
  and risk, 218
Glycogen storage disease Type 1B, 232
Gracilis flap operation, 194
Granulomas
  Crohn's disease diagnosis, 70
  Crohn's disease versus ulcerative colitis, 11–12
  as recurrence risk factor, 187
  transmission studies, 24
Gum hypertrophy, 40, 51–52

H2 blockers, 130
Hair loss, and sulfasalazine, 117
Hartmann pouch, 192, 250
Hay fever, 231
Head, physical examination, 51
Heart murmur, 53
Helper T-cells, 26
  suppressor cell ratio, 93
Hematuria, 42, 243
Hemolytic anemia, 234
Hemophilia, intramural bleeding, radiograph, 297
Hemorrhage. See also Bleeding
  and colectomy decision, 162
  universal ulcerative colitis, 34
Hepatic disease. See Liver disease
Hermansky–Pudlak syndrome, 233
Herniation, 191
Hinkle's pills, 251
Hip fistula, 209
HLA complex markers
  Crohn's disease versus ulcerative colitis, 11
  psoriasis, 230–231
  therapeutic implications, 27
HLA B27, 237–238
Hodgkin's disease
  versus Crohn's ileitis, 73–74
  radiograph, 303
Homosexual patients
  diagnostic problems, 81
  parasite diagnosis, 56
Hydrocortisone
  in Crohn's disease, 139
  in ulcerative colitis, 128–130
    risk factors, 129–130
Hydrocortisone enemas, 120, 127
Hydrocortisone foam, 126–127
Hydrogen breath test, 91
Hydronephrosis, 243
Hydroureter, 243
Hydroxychloroquine
  helper T-cell inhibition, 26
  therapeutic use, 124
Hyperoxaluria, 242

IgG antibodies, 10
Ileal–bladder fistula
   diagnosis and treatment, 208–209
   and massive hematuria, 42
Ileal cancer
   radiograph, 308
   surgery decision, 167
Ileal reservoir pouch. *See* Pelvic pouch
Ileitis
   differential diagnosis, 73–74, 84
   quality of life, 203
Ileoanal anastomosis, 173–174. *See also* Pelvic pouch
   advantages, 173–174
   cancer surveillance, 193
   in Crohn's disease of colon, 177–178
   stenosis complication, 185
Ileocecal fistula, 208
Ileocecal resection, 193–194
Ileojejunitis
   differential diagnosis, 73
   surgical choices, 176
Ileoproctectomy, 176
Ileorectal anastomosis, 173
   and cancer, 193, 224
   and Crohn's disease recurrence, 195
   indications for, 173
   results, 185
Ileosigmoid anastomosis, 177, 195
Ileosigmoid fistulas, 168, 208, 211
Ileosigmoidectomy, 176
Ileostomy, 171–172. *See also* Kock continent ileostomy
   choices for, 171–172
   in Crohn's disease of colon, 177, 192, 195–196
   complications, 192, 195–196
   and fistulas, 212
   patient selection, 65–66
   postoperative intestinal obstruction, 191–192
   quality of life, 203
   recurrence rates, 187–188, 195
   results, 181–182, 203
Ileostomy Society, 174
Ileovesicular fistulas, 209
Ileum
   in Crohn's disease, 43–44, 175–177
   and differential diagnosis, 73–74
   functional studies, 92

   operative choices, 175–177
Imaging techniques, 56–62
Immune complexes
   and extraintestinal disease, 236
   measurements of, 93–94
Immunologic status
   assessment, 93–94
   theories of, treatment implications, 25–27
Immunosuppressants
   available drugs, 115
   carcinogenic potential, 122–123
   in Crohn's disease, 140–141
   cutaneous fistula treatment, 213
   general considerations, 122–124
   maintenance use, 133, 142
   and pregnancy, 219
   in ulcerative colitis, 133
Impotence, 170–171, 182
Imuran®. *See* Azathioprine
Incubation period, Crohn's disease, 41
Indeterminate inflammatory disease, 9, 65–66
Infarction, differential diagnosis, 74–75
Inflammatory mediators, 24–25
Interleukin-2 production, 26
Intestinal obstruction
   Crohn's disease, 41
   following ileostomy, 191–192
   as immediate postoperative complication, 191
   and surgery, 166
   in ulcerative colitis, 34
Iritis, 35, 51, 237
Iron absorption, 91
Ischemia, small bowel, radiographs, 312, 313
Isoniazid, 23

Jejunal fistulas, 207
Jejunitis, skip areas, 176
Jejunum
   in Crohn's disease, 43–44, 175–177
   diagnostic errors, 76
   operative choices, 175–177
   perforation in Crohn's disease, 43–44
   skip lesions, 43, 176
Jews, genetic risk, 21, 218

Jöhne's disease, 23
Joint involvement, 35, 237–239

Kidney manifestations, 242–246
Kock continent ileostomy
　in Crohn's disease of colon, 177
　evaluation of, 171–172
　patient selection, 65–66, 172
　results, 182–183

Labeled leukocyte scans, 62
Labeled red cell scan, 61
Labial fistulas, 213
Labial infections, 40
Laboratory studies, 55, 88–90
Lactose intolerance, 146
Laxatives, 251
Left-sided colitis
　clinical presentation, 33–35
　drug therapy, 128
　prognosis, 98–100
Leukocyte scans, labeled, 62
Leukotriene inhibitors, 124
Lipomodulin, 25
5-Lipooxygenase inhibitors, 124
Liquid-formula diets, 147
Liver abscess, 61, 190–191
Liver disease, 246–248
　diagnosis, 247
　physical examination signs, 53
　radionuclear colloid scan, 61
　treatment, 247–248
Liver transplantation, 247–248
Lupus vasculitis, 311
Lymphangioma, radiograph, 300
Lymphocytes, 93
Lymphocytic colitis, 83
Lymphoid hyperplasia, radiograph, 314
Lymphoma, 73–74, 229
Lymphosarcoma, radiographs, 304, 305

Macrophage function, 232–233
Maintenance treatment
　Crohn's disease, 141–142
　general considerations, 112–113
　in ulcerative colitis, 131–133
Malabsorption, 241
　assessment, 91–92
　forms of, 241

Malignant melanoma, 229
Malnutrition
　assessment, 91–92
　causative factors, 146–147
Marital couples, contagion, 20
Massive hematuria, 42
Medical management. *See* Drug therapy
Megaloblastic anemia, 117
Melanoma, 229
Mepacrin, 82
6-Mercaptopurine
　in Crohn's disease, 140–142
　extraintestinal complications, 250
　fistula treatment, 208
　general considerations, 122–123
　maintenance use, 133, 142
　and pregnancy, 219
　and steroid weaning, 131
　in ulcerative colitis, 133
Mesalamine, 118–119
　in Crohn's disease, 134–135, 142
　general considerations, 118–119
　maintenance use, 133, 142
　in proctitis, 126
Mesenteritis, retractile, 76, 84
Meta-analysis, drug therapy, 154–158
Metastatic Crohn's disease, 40, 240
Methotrexate
　anti-inflammatory mechanism, 25
　maintenance use, 143
　therapeutic use, 123–124, 143
Metronidazole therapy
　and anerobic bacteria, 23–24
　in Crohn's disease, 121–122, 136–137, 143
　maintenance use, 143
　perirectal fistula treatment, 214
　side effects, 122, 136–137
Microscopic colitis
　diagnostic errors, 77
　differential diagnosis, 83
　histology, 83
Miscarriage, 218–219
Monozygotic twins, 21
Mucosal examination, 92
Mucosal stripping, 190
Multiple sclerosis, 231–232
Mycobacterial infection, 22–24
Myelodysplastic syndromes, 232
Myocarditis, 118

Neutrophil function, 232–233
Nursing mothers, 221
Nutritional management, 144–149
Nutritional status, assessment, 91–92

Obstruction, intestinal. *See* Intestinal obstruction
Olsalazine
  advantages, 118
  in Crohn's disease, 134–135
  and diarrhea, 135
  maintenance use, 133
  in proctitis, 126
Omega 3 fatty acids
  anti-inflammatory mechanism, 25
  in ulcerative colitis, 124
Orosomucoids, 89–90
Ovarian fistula, 209
Ovarian tubal mass, 40
Oxalate stones, 242

Pain, abdominal, assessment, 90–91
Pancolectomy. *See* Colectomy
Pancreatic fistulas, 209–210
Pancreatitis, 123, 140
Panophthalmitis, 237
Parental nutrition, 148–149
Parks procedure, 178–179, 214
Pelvic floor herniation, 191
Pelvic mass, 77
Pelvic pouch, 173–176. *See also* Pouchitis
  advantages, 173–176
  candidates for, 183
  in Crohn's disease of colon, 177–178
  quality of life, 183
  results, 182–185
Pentasa®, 118
Perforation
  and colectomy decision, 162–163
  in Crohn's disease, 38–39, 43–44, 47, 101–102
  and cutaneous fistulas, 213
  and jejunal stenosis, 43–44
  prognostic implications, 101–102
  in ulcerative colitis, 34
Perianal area
  fistulas, 213–215
  imaging techniques, 62
  physical examination, 52
Perianal disease, and surgery, 178–179

Peripheral joint manifestations, 237–239
Perirectal area
  examination of, 52, 215
  fistulas, 213–215
  imaging techniques, 62, 215
Perirectal disease, surgery, 179
Phantom rectum, 197
Physical examination, 51–53, 215
Placebo trials, 107–111
Plaquinal®, 25
Plasmacytoma of ileum, 73
  radiograph, 294
Platelet abnormalities, 90, 248
Polyps, and colectomy, 225–226
Postmenopausal women, 81
Postpartum period, 221
Pouchitis, 183–185, 198–199
  diagnostic implications, 66–67, 184–185
  and extraintestinal manifestations, 250
  incidence, 183, 198
  as man-made inflammation model, 183–184, 198
  severe form, 184–185, 198–199
  treatment, 184, 198–199
  in ulcerative colitis versus Crohn's disease, 11, 66–67, 184–185
Prednisolone metasulfabenzoate, 120
Prednisolone-21-phosphate, 120
Prednisone. *See* Steroids
Prednisone betasulfabenzoate, 121
Pregnancy, 217–221
  effect on underlying disease, 218–219
  and treatment, 219–221
Primary sclerosing cholangitis, 247
Proctitis, ulcerative, 31–33
  cancer risk, 97–98
  clinical presentation, 31–33
  drug strategies, 125–128
  prognosis, 97–98
Proctofoam®, 126–127
Proctoscopy, 58
Proctosigmoiditis, 125–128
Prognosis, 95
  Crohn's disease, 100–102, 110–111
  toxic dilation of colon, 102–103
  ulcerative colitis, 95–100, 109–110

Prophylactic colectomy, 163–164, 225
Pseudo-obstruction, radiograph, 310
"Pseudorheumatism," 131
Psoas spasm, 243
Psoriasis, 230–231
　HLA linkage, 230–231
　physical examination, 52
Psoriatic arthritis, 230–231
Psychoanalysis, 152
Psychosomatic factors, 36–37, 150
Psychotherapy, 150–153
Pulmonary embolism, 249
Pulmonary fibrosis, 117
Purinethol®. *See* 6-Mercaptopurine
Pyoderma gangrenosum, 239–240
　initial presentation, 35, 239
　recurrence of, 190
　surgery response, 165, 240
　treatment, 239–240
Pyrazinamide, 23

Quality of life, 200–203
　assessment issues, 200–202
　and cancer surveillance, 202
　Crohn's disease surgery, 202–203
　pelvic pouch operation, 183
　ulcerative colitis surgery, 181

Radiation enteritis, 74
　radiograph, 296
Radiation exposure, 57, 74, 220
　acute myelogenous leukemia, 232
　in pregnancy, 220
Radiography
　differential diagnosis problems, 73–74
　indications for, 56–57
Radionuclear colloid scan, 61
"Reactive" arthritis, 197, 237–238
Recovery rate, ulcerative colitis, 97, 99
Rectal anastomosis, 177
Rectal bleeding
　and differential diagnosis, 75, 80–81
　ulcerative proctitis, 32–33
　universal ulcerative colitis, 33–34
Rectal cancer, 98
Rectal pain, 215
Rectal-sparing operations, 172–175

Rectosigmoid colon surveillance, 192–193, 226
Rectovaginal fistula, 179
Rectum
　cancer surveillance, 192–193, 226
　physical examination, 52, 215
　solitary ulcer, 80–81
Recurrence
　and anastomosis, Crohn's disease, 176, 187–188, 194–195
　clinical versus endoscopic, 186–187
　Crohn's disease, 101–102, 186–188, 194–195
　and surgery, 186–188
　toxic dilation of colon, 103
　ulcerative colitis, 98–99
Red cell scan, labeled, 61
Refractory pouchitis, 66–67
Relaxation techniques, 153
Remission
　Crohn's disease, 100–101, 111
　prediction of, 109
　in ulcerative colitis, 96–97, 99, 109–110
Renal amyloidosis, 243–246
　clinical presentation, 245
　colchicine in, 245–246
　treatment, 245–246
Renal manifestations, 118, 242–246
Reoperation, 186–188
Reticuloendothelial malignancies, 228–229
Retractile mesenteritis, 76, 84
Rheumatoid arthritis, 238
Riedel's struma of thyroid, 84
Rifampicin, 23
Rifaputin, 23
Rowasa® enemas, 119, 125

Sacroiliitis, 237
Saddle bags, 43
Schilling test, 91
Scirrhous carcinoma, radiograph, 308
Sclerosing cholangitis
　colectomy effect on, 165
　diagnosis and treatment, 242–243
　incidence, 234
Segmental colitis, 68, 73
Self-inflicted trauma, 80–81, 250
Self-limited colitis, 107–108
Sepsis, Crohn's disease, 42

Serositis, 192
Serum albumin, 90–91
Sex factors
  pouchitis, 198
  and recurrence, 187
Short-chain fatty-acid enemas, 196
Siblings, 21
Sigmoid diverticulitis, 75, 78–79
Sigmoidoscopy, 54
  contraindications, 58–59
  in ulcerative colitis, 58–59
Silent Crohn's disease, 40–41
Skin manifestations, 52, 239–240
Skin reactivity, 93
Skip lesions, 43, 176
Smoking
  cessation of, 35
  and risk, 19–20
  in twins, 21
Solitary ulcer of rectum, 80–81
Spastic colon, 81
Sperm count, 117, 218
Sphincter muscles, 214–215
Splenomegaly, 53
Squamous cell cancer, 228
Stenosis
  and ileoanal anastomosis, 185
  ileosigmoid fistula pathogenesis, 208
  operative choices, Crohn's disease, 175–176
  and jejunal perforation, 43–44
  and retractile mesenteritis, 84
Steroids
  acrylic resin coatings, 121
  anti-inflammatory mechanism, 25–27
  arthralgia cause, 131
  available drugs, 114
  in Crohn's disease, 137–140
    dosage, 138
  general considerations, 119–121
  intravenous use, 127–130
    risk factors, 129–130
  long-term use, dangers of, 132, 138–139
  meta-analysis, 156
  nonabsorbable, 120–121
  and pregnancy, 219–221
  in proctitis, 127–128
  in severe ulcerative colitis, 95–96
  side effects, 132, 137–138
  topical, 120–121
  in universal ulcerative colitis, 128
  weaning, 127–128, 131, 138–140
Streptomycin, 23
Strictures of colon
  and cancer, 223–224
  colectomy indication, 164
  current diagnostic practices, 223–224
Stricturoplasty, 44, 176
Stroke, 249
Subtotal colectomy
  cancer surveillance, 226
  complications, 189–190, 192–193
Sugar-free diets, 148
Sugar-rich diet, 22
Sulfasalazine, 116–118
  anti-inflammatory mechanism, 25, 27
  clinical usefulness, 116
  in Crohn's disease, 134–135
  intolerance to, 126
  maintenance therapy value, 27, 132–133
  meta-analysis, 156
  and pregnancy, 219
  in proctitis, 126
  side effects, 116–118
  and spermatogenesis, 117, 218
Sulfathaladine, 122
Suppressor T-cells, 26, 93
Surgery, 161–199
  choices, 170–179
  Crohn's disease, 165–169, 175–179, 186–188
  decision for, 161–169
  and diagnostic problems, ileum, 79
  elective, 164–165, 168–169
  extraintestinal manifestations, 250
  postoperative problems, 189–199
  prediction of, ulcerative colitis, 100
  and pregnancy, 221
  results, 180–188
  in ulcerative colitis, 161–165, 170–175

T-cells
  helper to suppressor ratio, 93
  immunologic theories, 26
Thrombocyte count, 90
Thrombocytopenic purpura, 118

Thrombocytosis, 40
Thrombosis, and platelet count, 90, 244
Tissue necrosis factor, 93
Toes
 gangrene, 77, 248–249
 physical examination, 52
Total parenteral nutrition, 148–149
Toxic dilation of colon
 and colectomy decision, 163
 physical examination indicators, 53
 prognosis indicator, 102–103
 in universal ulcerative colitis, 34
Toxicortical pivalate, 120–121
Tuberculosis
 Crohn's disease association, 22–23, 73
 radiographs, 298, 299
Turner's syndrome, 231
Twin studies, 12–13, 21

Ulcerative colitis
 and cancer, 222–224
  diagnostic errors, 75–76, 223–224
  endoscopy timing, 59
  and ileorectal anastomosis, 185, 193
  proctitis risk, 97–98
  prognosis, 224
  risk, 222–223
  subtotal colectomy risk, 192–193
 clinical assessment, 88–89
 clinical presentation, 31–47
 Crohn's disease distinction, 8–14, 65–67
 diagnostic difficulties, 68–69
 diagnostic errors, 75–76
 drug strategies, 125–133, 154–158
  meta-analysis, 154–158
 endoscopy, 58–59
 epidemiologic trends, 13
 etiologic theories, treatment implications, 19–27
 genetic factors, 20–22
 immunologic etiologies, and treatment, 25–27
 infectious agents, 22–24

 maintenance drug therapy, 131–133
 natural history of, 11, 107–109
 pathology, 11–12
 prognosis, 95–100, 109–110
 psychological factors, 152
 quality of life, 202
 severe form, prognosis, 95–97
 surgery, 161–165, 170–175, 181–186
  choices, 170–175
  complications, 192–193
  decision for, 161–165
  results, 181–186
Ulcerative proctitis. See Proctitis, ulcerative
Ultrasound, 61
Umbilical fistula, 211
Umbilical infection, 40
Universal ulcerative colitis
 clinical presentation, 33–35
 drug therapy, 128
 prognosis, 98–100
Ureter fistulas, 209
Uric-acid stones, 242
Ursodecholic acid, 247

Vaginal fistulas, 209–210
Vascular complications, 248–249
Venous thrombosis, 90, 244
Vitamin B12 malabsorption, 91–92, 241
Vitamin B12 supplements, 146
Vitamin D supplements, 146
von Willebrand's factor, 248

Walled-off perforation
 abdominal wall, 211–212
 and colectomy decision, 163
 in ulcerative colitis, 34
Weaning from steroids, 127–128, 131, 138–140
Whipple's disease, 79

Yersinia enterocolitica, 46
Yersinia infections, 46, 73–74

Zileuton, 124
Zinc, 146
Zollinger–Ellison syndrome, 76